Andreas Riegler

Ferromagnetic resonance study of the Half-Heusler alloy NiMnSb

Andreas Riegler

Ferromagnetic resonance study of the Half-Heusler alloy NiMnSb

The benefit of using NiMnSb as a ferromagnetic layer in pseudo-spin-valve based spin-torque oscillators

Südwestdeutscher Verlag für Hochschulschriften

Impressum/Imprint (nur für Deutschland/only for Germany)
Bibliografische Information der Deutschen Nationalbibliothek: Die Deutsche Nationalbibliothek verzeichnet diese Publikation in der Deutschen Nationalbibliografie; detaillierte bibliografische Daten sind im Internet über http://dnb.d-nb.de abrufbar.
Alle in diesem Buch genannten Marken und Produktnamen unterliegen warenzeichen-, marken- oder patentrechtlichem Schutz bzw. sind Warenzeichen oder eingetragene Warenzeichen der jeweiligen Inhaber. Die Wiedergabe von Marken, Produktnamen, Gebrauchsnamen, Handelsnamen, Warenbezeichnungen u.s.w. in diesem Werk berechtigt auch ohne besondere Kennzeichnung nicht zu der Annahme, dass solche Namen im Sinne der Warenzeichen- und Markenschutzgesetzgebung als frei zu betrachten wären und daher von jedermann benutzt werden dürften.

Coverbild: www.ingimage.com

Verlag: Südwestdeutscher Verlag für Hochschulschriften GmbH & Co. KG
Heinrich-Böcking-Str. 6-8, 66121 Saarbrücken, Deutschland
Telefon +49 681 37 20 271-1, Telefax +49 681 37 20 271-0
Email: info@svh-verlag.de

Approved by: Würuzburg, Julius-Maximilians-Universität, Dissertation, 2011

Herstellung in Deutschland:
Schaltungsdienst Lange o.H.G., Berlin
Books on Demand GmbH, Norderstedt
Reha GmbH, Saarbrücken
Amazon Distribution GmbH, Leipzig
ISBN: 978-3-8381-3136-8

Imprint (only for USA, GB)
Bibliographic information published by the Deutsche Nationalbibliothek: The Deutsche Nationalbibliothek lists this publication in the Deutsche Nationalbibliografie; detailed bibliographic data are available in the Internet at http://dnb.d-nb.de.
Any brand names and product names mentioned in this book are subject to trademark, brand or patent protection and are trademarks or registered trademarks of their respective holders. The use of brand names, product names, common names, trade names, product descriptions etc. even without a particular marking in this works is in no way to be construed to mean that such names may be regarded as unrestricted in respect of trademark and brand protection legislation and could thus be used by anyone.

Cover image: www.ingimage.com

Publisher: Südwestdeutscher Verlag für Hochschulschriften GmbH & Co. KG
Heinrich-Böcking-Str. 6-8, 66121 Saarbrücken, Germany
Phone +49 681 37 20 271-1, Fax +49 681 37 20 271-0
Email: info@svh-verlag.de

Printed in the U.S.A.
Printed in the U.K. by (see last page)
ISBN: 978-3-8381-3136-8

Copyright © 2012 by the author and Südwestdeutscher Verlag für Hochschulschriften GmbH & Co. KG and licensors
All rights reserved. Saarbrücken 2012

Parts of this thesis have been published elsewhere; other manuscripts are in preparation:

- A. Riegler, F. Lochner, C. Gould, G. Schmidt, and L. W. Molenkamp, *Substrate induced uniaxial anisotropy in (111)NiMnSb layers*, yet unpublished

- A.Riegler, F. Lochner, C. Gould, G. Schmidt and L. W. Molenkamp, *Magnetostatic modes in arrays of sub-micrometer sized elements of (001)NiMnSb*, yet unpublished

- A.Riegler, F. Lochner, C. Gould, G. Schmidt and L. W. Molenkamp, *Very high q-factor for spin-torque oscillators in zero field*, yet unpublished

- G.de Loubens, A. Riegler, B. Pigeau, F. Lochner, F. Boust, K. Y. Guslienko, H. Hurdequint, L. W. Molenkamp, G. Schmidt, A. N. Slavin, V. S. Tiberkevich, N. Vukadinovic, and O. Klein,*Bistability of vortex core dynamics in a single perpendicularly magnetized nanodisk*, Physical Review Letters, 102**17**:177602, May 2009

- B. Pigeau, G. de Loubens, O. Klein, A. Riegler, F. Lochner, G. Schmidt, L. W. Molenkamp, V. S. Tiberkevich, and A. N. Slavin,*A frequency-controlled magnetic vortex memory*, Applied Physics Letters, 96**13**:132506, March 2010

- B. Pigeau, G. de Loubens, O. Klein, A. Riegler, F. Lochner, G. Schmidt, L. W. Molenkamp, *Optimal control of vortex core polarity by resonant microwave pulses*, Nature, Vol. DOI: 10.1038/nphys1810, 2010

Contents

Zusammenfassung		1
Summary		3
1 Introduction		5
2 Theory		7
2.1	Energy consideration	9
	2.1.1 shape anisotropy	9
	2.1.2 Zeeman energy	9
	2.1.3 crystal anisotropy	9
2.2	Magnetization dynamics	12
2.3	Ferromagnetic resonance (FMR)	13
	2.3.1 resonance condition	14
	2.3.2 damping mechanisms	16
2.4	Concept of spin pumping and spin sinking	19
2.5	Giant magnetoresistance (GMR)	21
	2.5.1 Mott two spin channel-model	21
	2.5.2 intrinsic GMR	22
	2.5.3 extrinsic GMR	23
2.6	Spin-torque	25
3 Samples		29
3.1	The half-Heusler alloy NiMnSb	31
	3.1.1 crystal structure of NiMnSb	32
	3.1.2 growth of NiMnSb	32
3.2	Different types of spin-valves	33
	3.2.1 Spin-valves	33
	3.2.2 Pseudo-spin-valve	34
4 Results		37
4.1	Ferromagnetic resonance spectrometer	39
4.2	Ferromagnetic resonance measurements on NiMnSb	41
	4.2.1 NiMnSb grown on 001 InP substrate	41

		4.2.2	NiMnSb grown on 111 InP substrate	49
	4.3	Spin pumping		55
	4.4	Magnetostatic modes		61
	4.5	Summary of hf measurements of NiMnSb		70
	4.6	Spin-torque devices using NiMnSb		71
		4.6.1	magnetoresistance	71
		4.6.2	current induced switching	72
		4.6.3	spin-torque oscillator	74

5   Conclusion and outlook   85

A   Appendices   87
   A.1   Scattering   87
   A.2   Sample fabrication   89

Bibliography   98

List of Figures

2.1 Orientation of the magnetization and the external magnetic field 10
2.2 Coordinate system which is used to describe motion of the magnetization and ferromagnetic resonance 12
2.3 a) magnetization motion without damping; b) motion of the magnetization with damping, blue arrow indicating the damping torque which is oriented onto the direction of the effective field 17
2.4 A ferromagnetic layer (F) sandwiched between two paramagnet electron reservoirs (NM) with equal energy (being in thermal equilibrium) 19
2.5 schematic giant magnetoresistance for a) parallel and b) antiparallel alignment 21
2.6 Band structure of a ferromagnet in an external field, the external field is parallel to the spin ↑ and therefore the energy is lowered by an amount . . 23
2.7 Scattering for antiparallel alignment (a) and for parallel alignment (b) . . . 24
2.8 Coulomb and exchange potentials versus position ξ in a five layer system is shown whilst the particle flow is rightward the charged flow is leftward. Vector diagram shows the spin moments $S_{1,2}$ and their current driven velocities $\dot{S}_{1,2}$ (from [Slon 96]) 25
2.9 An electron flux from the free to the fixed layer results in a torque on the free layer which points away from the fixed layer resulting in an antiparallel alignment (a) while an electron flux in the opposite direction results in a favoured parallel alignment due to direction of the resulting torque (b) . . 27

3.1 $C1_b$ structure of NiMnSb (red:Ni, blue:Mn and green:Sb) 31
3.2 a) Magnetoresistance and b) magnetization curve for a spin-valve structure with antiferromagnet (AFM) to 'pin' the fixed layer (the curves show field dependent measurement in which the field is lower than the exchange field of the antiferromagnet), in c) a typical heterostructure is shown 34
3.3 a) magnetization minor loop for spin valve with antiferromagnet and b) complete loop hysteresis; in region IV the field is strong enough to align the antiferromagnet and therefore adding an extra magnetic moment to the magnetization 34

3.4 magnetoresistance and magnetization curve for a pseudo-spin-valve structure. The difference in the switching fields can be achieved by either ferromagnet with different coercive fields or by engineering an additional shape anisotropy 35

4.1 Schematic of the used fmr spectrometer: the signal is divided using a splitter/combiner in a reference arm and a signal arm which are afterwards recombined by using a second splitter/combiner; the RF-amplifier is a broad-band amplifier with a gain of 43 dB; the modulation frequency is 156 Hz 39

4.2 A typical spectrum is shown (a) derivative as measured with lock-in technique; b) direct signal); the intensity fluctuations are due to the rotating of the external field and therefore having only for 0° and 180° a perpendicular orientation of the static field and the alternating GHz field; the graphs are recorded using different sensitivities of the lock-in 41

4.3 resonance field versus in-plane angle for 10 nm NiMnSb for three different frequencies (black:8 GHz, red:12,5 GHz and green:14 GHz) and fits using parameters shown in tab. 4.1 42

4.4 Resonance field versus in-plane angle for 8 GHz (black), 12.5 GHz (red) and 14 GHz (green) (open boxes) and fitted angle dependencies (solid lines) using parameters from table 4.1 43

4.5 The solid line is the analytically found fit (parameters according to tab. 4.1) and the open boxes are representing the measurements for three frequencies (black:8 GHz, red:12.5 GHz and green:14 GHz) 44

4.6 Effective magnetization over inverse thickness is plotted. The decrease in the magnetization with decreasing thickness is consistent with the definition of the effective magnetization as the sum of the saturation magnetization and the out-of-plane uniaxial anisotropy which scales inverse to the thickness (see equ. 2.30) 45

4.7 The figure show the in-plane uniaxial anisotropy field strength versus the inverse thickness. A clear linear dependency (solid line corresponds to fit) indicates the expected 1/d dependency 46

4.8 In a) the damping of the 10 nm thick NiMnSb layer is shown and in b) the inhomogeneous broadening (the zero frequency offset) is shown 46

4.9 On the left hand side (a) the damping is shown in a polar coordinate system whilst on the right hand side (b) the non-frequency offset (inhomogeneous broadening) of the 20 nm (001)NiMnSb layer is shown 47

4.10 Damping and inhomogeneous broadening extracted from the line width measurements of the 40 nm NiMnSb layer 47

4.11 Plot shows the line width over ω/γ for the 40 nm NiMnSb layer, the slope representing the damping, namely 4.74×10^{-3} along the (100) direction and 3.33×10^{-3} along the $(1\bar{1}0)$ direction 48

- 4.12 a) Picture of an atomic force microscope (AFM) scan of the surface of a 10 nm thick NiMnSb layer grown on (111)(In,Ga)As buffer on (111)b InP substrates; b) drawing of the (111)NiMnSb sample with its triangular shape . . . 49
- 4.13 Resonance field versus angle for a 10 nm thick NiMnSb layer grown on (111)(In,Ga)As buffer, a pronounced uniaxial anisotropy is visible with its easy axis along the $(\overline{1}\overline{1}2)$ crystal direction 50
- 4.14 An angle dependency (resonance field versus in-plane angle with respect to the $(2\overline{1}\overline{1})$ crystal direction) of 20 nm NiMnSb layer grown on (111) InP is shown 50
- 4.15 Angle dependent measurement of the resonance field (40nm (111)NiMnSb) with a pronounced uniaxial anisotropy similar to all previously measured NiMnSb layers grown on (111)(In,Ga)As 51
- 4.16 Anisotropy field strength over layer thickness (a) and over inverse layer thickness (b) is shown. The red line is a linear fit, which fits adequately well to both datasets 52
- 4.17 Line width of the absorption peaks for the angle dependent in-plane FMR measurement of the NiMnSb layers which are grown on (111)(In,Ga)As, a) 20 nm thickness and b) 40 nm thickness 52
- 4.18 Damping parameter α for different capping materials on NiMnSb with different thicknesses (✶:Cu, ▲:Pt, ▼:Ta, ●:MgO, ■:Ru);a) damping over thickness;b) damping over the inverse thickness for all capping materials . 55
- 4.19 Pumping using MgO as capping layer; the increase with increasing thickness is contradicting the spin-pumping concept 55
- 4.20 NiMnSb with Cu as capping layer comparable to MgO 56
- 4.21 Damping over the inverse thickness of the NiMnSb layer for Ta, Pt and Ru (▲:Pt ,▼:Ta, ■:Ru) as capping metals 57
- 4.22 the figure above shows the damping parameter over the inverse-thickness (lower x-axis; over the thickness, upper x-axis) for NiMnSb layer capped with Ru 57
- 4.23 a) shows a the damping of a Permalloy (Py) layer covered with Ta for different thicknesses, the linear fit results in a slope of 0.0534 while in b) NiMnSb layer capped with Ta is shown and the corresponding best linear fit having a slope of 0.035 58
- 4.24 The figure above shows a comparison of the damping parameters for Py and NiMnSb layers covered with Ta (▲:NiMnSb, ■:Py) 58
- 4.25 Damping of the NiMnSb layer capped with 30 nm Pt over inverse thickness 59
- 4.26 Line width of absorption measured for rectangular NiMnSb elements (600 times 1200 nm^2, 40 nm thick) with the external field applied parallel to the long side of the stripes, the solid red line indicates the best linear fit and therefore results in a damping α of 2.3×10^{-3} 61
- 4.27 The plots show the spectra of a) two 10 time 10 μm^2 squares and b) an array of 10 times 10 discs with a diameter 2 μm 62

4.28 Magnetostatic modes for arrays of stripes with different aspect ratio: a) 600 x 1200 nm^2, b) 600 x 2000 nm^2 and c) 600 x 3400 nm^2 stripes 62

4.29 Comparison of spectra for stripes with different aspect ratio (curves are offset by 1 for clarity) 63

4.30 Above an illustration of the used pulse (a) which is used in the simulation to start the relaxation process in the elements and (b) the resulting, in time decaying, magnetization of the element is shown 64

4.31 Measurement and FFT of the simulated data for the 10×10 μm^2 squares . 65

4.32 For the discs (measurement upper graph and FFT of simulation lower graph) the resonances correspond well 66

4.33 The upper graph shows the measured spectrum of an array of 10 times 10 600×1200 nm^2 stripes while the lower graph shows the corresponding FFT of the simulation 67

4.34 The lower graph show the FFT of the OOMMF simulation of an array of 10 times 10 stripes with dimensions of 600×2000 nm^2 and is in very good agreement to the measurement of the FMR as can be seen in the upper plot 68

4.35 An array of 10 times 10 stripes (600×3400 nm^2) is measured using FMR (upper graph) and compared to simulations done with OOMMF (lower plot) 69

4.36 Room temperature magnetoresistance measurement of a NiMnSb based pseudo-spin-valve ((40 nm)NiMnSb/(10)Cu/(6)Py) measured in an external field swept along the long axis of the elliptically shaped pillar (80 x 200 nm^2) 71

4.37 Magnetoresistance measurement for the same device done at low temperature (80 × 200 nm^2; (40)NiMnSb/(10)Cu/(6)Py) 73

4.38 Room temperature current induced switching measurement of a elliptical shaped (80 × 200 nm^2) NiMnSb based pseudo-spin-valve with the same composition ((40)NiMnSb/(10)Cu/(6)Py) as above 74

4.39 Low temperature measurement of the same device presented above offers an expected decrease in current densities which are necessary to change the magnetic configuration 75

4.40 Preliminary magnetoresistance measurement of the STO device (elliptical shaped (100 × 200 nm^2; (40)NiMnSb/(10)Cu/(6)Py) to verify the DC-characteristics 75

4.41 Resistance versus current is plotted. Two well separated states are visible(high resistance, antiparallel alignment, low resistance, parallel alignment) and the corresponding current densities of 1.79×10^7 A/cm^2 for switching from anti- to parallel and 2.3×10^7 A/cm^2 to reverse the configuration. . 76

4.42 Schematics of the measurement of spin-torque oscillators 77

4.43 Noise figure of the measurement setup, measured with the tips lowered on the sample but without an applied current 78

4.44 Illustrated procedure to prepare the device in anti-parallel magnetic configuration before applying current and measure spin-torque oscillations . . 79

4.45	Spectrum for the pseudo-spin-valve ((40)NiMnSb/(10)Cu/(6)Py) with an applied current of -2.983 mA	79
4.46	Spectrum for the device measured in an external field of 207 Oe for different currents (a) and the corresponding peak position and q-factors over applied currents (b)	80
4.47	Spectrum for different currents (black:-3.128 mA; red:-3.2021 mA; green:-3.2759 mA; blue:-3.3452 mA; cyan:-3.4207 mA; magenta:-3.4935 mA) . . .	80
4.48	Measurement of the spectrum of the same device in an external applied magnetic field of 169 Oe (a) and the current dependencies for q-factor and peak position is shown in b)	81
4.49	Spectrum for the oscillator in an external field of 135 Oe (a) and in b) the corresponding current dependent peak position and q-factors are shown . .	81
4.50	Peak position of the first modes of the device over the applied external field is shown, higher fields (207 and 169 Oe) show two modes while the spectrum for lower fields (135 and 0 Oe) show only one mode which is likely a distinct mode due to the fact that they show not the same field dependency as the measured modes for higher fields	82
4.51	Emitted high frequency of the device operating in the absence of an applied external magnetic field (a) and the frequency and q-factor over applied currents (b) are shown	82
A.1	schematics of the spin dependent scattering	87
A.2	process starts from the unpatterned heterostructure (a) to the final device for DC characterization (i) and for AC measurements (j)	90

Zusammenfassung

Seit der Entdeckung des Spin-Torque durch Berger und Slonczewsky [Berg 96, Slon 96] im Jahre 1996 gewann dieser Effekt immer mehr an Einfluss in dem Gebiet der Spintronic. Dies geschah besonders durch den Einfluss des Spin-Torque auf die Informationsspeicher- und Kommunikationstechnologien (z.B. die Möglichkeit einen magnetischen Zustand eines Speicherelementes mit Hilfe von Strom und nicht wie bisher durch das Anlegen eines magnetischen Feldes zu ändern, oder die Realisierung eines hochfrequenten Spin-Torque-Oszillator (STO) [Tsoi 00]). Aufgrund des direkten Zusammenhangs zwischen der Dämpfung in Ferromagneten und der kritischen Stromdichte, die nötig ist um ein Spin-Ventil zu schalten oder ein Präzidieren der Magnetisierung zu induzieren, wurde die Forschung an Ferromagneten mit geringer Dämpfung zunehmend forciert.

In dieser Arbeit werden Studien der ferromagnetischen Resonanz (FMR) von NiMnSb Schichten und Transportmessungen an NiMnSb basierten Spin-Ventilen präsentiert. Das Halbmetall NiMnSb ist mit einer theoretischen 100%igen Spinpolarisation [DeGr 83] prädestiniert für die Verwendung in GMR [Baib 88, Bina 89] Elementen. Neben der theoretisch vorhergesagten hohen Spinpolarisation zeigen die durchgeführten FMR Messungen einen überaus geringen Dämpfungsfaktor α für dieses Material. Dieser liegt in der Größenordnung von wenigen 10^{-3}. Somit ist die Dämpfung in NiMnSb um den Faktor zwei geringer als in Permalloy und gut vergleichbar mit epitaktisch gewachsenen Eisen-Schichten. Neben den guten Dämpfungseigenschaften zeigen jedoch theoretische Modelle den Verlust der 100%igen Spinpolarisation durch das Brechen der Translationssymmetrie an Grenzflächen und das Kollabieren der Aufspaltung im Minoritäts-Spin-Band. Da ein Wachstum in (111) Richtung diesen Prozess entgegen wirken kann [Wijs 01], werden in dieser Arbeit zudem auf (111)(In,Ga)As gewachsene NiMnSb Schichten mittels FMR untersucht. Die Messungen an diesen Proben zeigen, im Vergleich zu (001) orientierten Schichten, eine erhöhte Dämpfung. Zudem kann bei diesen Schichten eine schichtdickenabhängige uni-direktionale magnetische Anisotropie gemessen werden [Rieg 10b].

Im Hinblick auf den möglichen industriellen Einsatz in Speicherelementen werden überdies Messungen an Sub-Mikrometer großen NiMnSb Elementen auf (001) orientierten Substraten präsentiert. Die Elemente wurden mittels Elektronenstrahllithografie hergestellt und mittels FMR vermessen. Auch die so prozessierten Schichten zeigen einen Dämpfungsfaktor im unteren 10^{-3} Bereich. Das Auftreten von magnetostatischen Moden in den Messungen [Rieg 10a] ist ein weiterer indirekter Nachweis der hohen Qualität der NiMnSb-Schichten.

Im Jahre 2001 wurde von Mizukamie und seinen Kollegen [Mizu 01a, Mizu 01b] eine dickenabhängige Erhöhung der Gilbertdämpfung bei, mit Metallen bedeckten, Permalloy-Schichten beobachtet. Im Jahr darauf wurde von Tserkovnyak, Brataas und Bauer eine Theorie erarbeitet die dieses Phänomen auf ein Pumpen von Spins aus dem Ferromagneten in die Metalschicht zurückführt [Tser 02]. Aus diesem Grund werden Messungen von NiMnSb Schichten, die mit verschiedenen Metallen und Isolatoren in-situ vor Oxidation geschützt wurden, präsentiert.

Nach diesen materialspezifischen Voruntersuchungen werden auf NiMnSb und Permalloy basierte Pseudo-Spin-Ventile unter Verwendung eines selbst ausrichtenden lithografischen Prozesses hergestellt [Gall 97]. Transportmessungen an den Proben zeigen ein GMR-Verhältnis von 3,4% bei Raumtemperatur und fast das doppelte bei tiefen Temperaturen. Diese sind sehr gut vergleichbar mit den besten veröffentlichten GMR-Verhältnissen für Einzelschichtsysteme. Überdies kann in den Experimenten eine viel versprechend geringe kritische Stromdichte, die nötig ist, um die magnetische Orientierung zu ändern, gemessen werden. Diese ist vergleichbar mit kritischen Stromdichten aktuellster metallbasierter GMR-Elemente [Fert 04] oder auf dem Tunneleffekt basierenden Spin-Ventilen [Huai 04].

Das eigentliche Potential der auf NiMnSb basierenden Spin-Ventile wird erst ersichtlich wenn diese als STO zum Emittieren hochfrequenter, durchstimmbarer und schmalbandiger elektromagnetischer Wellen verwendet werden. Auf Heusler basierende STO zeigen einen überdurchschnittlich hohen q-Faktor von 4180, sogar im Betrieb ohne extern angelegtes Magnetfeld [Rieg 10c]. Dieser ist um mehr als eine Größenordnung höher als der höchste veröffentliche q-Faktor eines ohne externes Feld arbeitenden STO [Devo 07]. Während die Heusler basierten STO ebenso wie alle anderen STO unter einer geringen Ausgangsleistung leiden, machen die Maßstäbe im Sub-Mikrometer Bereich eine On-Chip Herstellung möglich. Somit kann durch ein Parallelschalten von gekoppelten Oszillatoren eine Erhöhung der Ausgangsleistung erzielt werden [Manc 05].

Summary

Since the discovery of spin torque in 1996, independently by Berger and Slonczewski, [Berg 96, Slon 96] and given its potential impact on information storage and communication technologies, (e.g. through the possibility of switching the magnetic configuration of a bit by current instead of a magnetic field, or the realization of high frequency spin torque oscillators (STO) [Tsoi 00]), this effect has been an important field of spintronics research. One aspect of this research focuses on ferromagnets with low damping. The lower the damping in a ferromagnet, the lower the critical current that is needed to induce switching of a spin valve or induce precession of its magnetization.

In this thesis ferromagnetic resonance (FMR) studies of NiMnSb layers are presented along with experimental studies on various spin-torque (ST) devices using NiMnSb. NiMnSb, when crystallized in the half-Heusler structure, is a half-metal which is predicted to have 100% spin polarization [DeGr 83], a consideration which further increases its potential as a candidate for memory devices based on the giant magnetoresistance (GMR) effect [Baib 88, Bina 89]. The FMR measurements show an outstandingly low damping factor α for NiMnSb, in low 10^{-3} range. This is about a factor of two lower than permalloy and well comparable to lowest damping for iron grown by molecular beam epitaxy (MBE). According to theory the 100% spin polarization properties of the bulk disappear at interfaces where the break in translational symmetry causes the gap in the minority spin band to collapse but can remain in other crystal symmetries such as (111) [Wijs 01]. Consequently NiMnSb layers on (111)(In,Ga)As buffer are characterized in respect of anisotropies and damping. The FMR measurements on these samples indicates a higher damping that for the 001 samples, and find a thickness dependent uniaxial in-plane anisotropy [Rieg 10b].

Investigations of the material for device use is pursued by considering sub-micrometer sized elements of NiMnSb on 001 substrates, which were fabricated by electron-beam lithography and measured by ferromagnetic resonance. The damping remains in the low 10^{-3} range as determined directly by extracting the Gilbert damping from the line width. Additionally magnetostatic modes are observed in arrays of elements, which is further evidence of high material quality of the samples [Rieg 10a].

By sputtering various metals on top of the NiMnSb, spin pumping [Tser 02] from the ferromagnet into the non-magnetic layer is investigated.

After these material investigations, pseudo-spin-valves using NiMnSb as one of the ferromagnet, in combination with Permalloy were fabricating using a self-aligned lithography process[Gall 97]. These samples show a GMR ratio of 3.4% at room temperature and almost double at low temperature, comparing favourably to the best single stack GMR structures reported to date.

Moreover, current induced switching measurements show promisingly low current densities are necessary to change the magnetic orientation of the free layer. These current densities compete with state-of-the-art GMR devices [Fert 04] for metal based structures and almost with tunnel junction devices [Huai 04]).

The true potential of these devices however comes to light when they are operated as spin torque oscillators to emit high frequency, tunable, narrow spectrum electromagnetic waves. These Heusler based STOs show an outstanding q-factor of 4180, even when operating in the absence of an external field [Rieg 10c], a value which bests the highest value in the literature by more than an order of magnitude [Devo 07].

While these devices currently still suffer from the same limited output power as all STO reported to date, their sub-micron lateral dimensions make the fabrication of an on-chip array of coupled oscillators [Manc 05], which is a promising path forward towards industrially relevant output power.

Chapter 1

Introduction

During the last years a lot of research has been done concerning the high frequency characteristics of ferromagnets. With the discovery of spin-torque [Berg 96, Slon 96] in ferromagnetic spin-valves and pseudo-spin-valves the interest increased. In spin-torque devices the damping of the ferromagnets plays a major role due to a direct relation between damping and threshold current densities which are necessary to switch the magnetic formation within spin-valves. Damping parameters of ferromagnets are as low as 6 to 7×10^{-3} [Koba 09, Kala 06] for Py or 4×10^{-3} for MBE grown Fe [Urba 01]. Besides high frequency characteristics of such ferromagnets the spin polarization is another important parameter. The low damping is necessary for switching the magnetic configuration by passing a current while the polarization gives rise to the GMR ratio and therefore to a possible industrial interest. One good candidate (which is predicted to show high spin-polarization) is the class of half-metals which are predicted to be 100% spin polarized [DeGr 83]. The present work has its focus on the half-Heusler alloy NiMnSb [Van 00, Van 01] which exhibits a low damping parameter α in the order of 3 to 4×10^{-3} in combination with a curie temperature of over 700 K. The low damping combined with a high spin polarization are qualifying factors for a possible use of this material in spin-torque as well as in GMR devices.

In chapter two the theory necessary for this thesis is presented, namely the internal energies of a ferromagnet which are the origin of anisotropies followed by magnetization dynamics and at last the giant magnetoresistance and spin-torque concept. The chapter starts with the theory necessary for ferromagnetic resonance. After describing different mechanism involved in the motion of magnetization in a ferromagnetic layer a concept of spin pumping and spin sinking is presented. Afterwards the last part of this chapter is focused on GMR and ST concepts.

Chapter three is introducing NiMnSb to the reader. Its crystal structure and a short overview of the growth procedure is given. This is followed by an introduction to different spin-valves and its characteristics. In chapter four ferromagnetic resonance measurements of grown (001)- and (111)-oriented NiMnSb layers are presented. The anisotropies and damping of layers are characterized by these measurements. Compared to preliminary

FMR measurements [Kove 05] performed on similar layers which were also grown at the Experimentelle Physik III department of the University of Wuerzburg, the latest batch of NiMnSb layers shows an even lower damping after changed growth conditions (namely the changed growth conditions to work under the melting point of Ni, see Florian Lochners PhD for more information: [Loch 10b]). Afterwards FMR measurements of (001)NiMnSb layers gapped using different non-magnetic metals are presented. Spin pumping from the ferromagnetic NiMnSb layer into the metal is measured by thickness dependent damping measurements. In the last part of the chapter the focus is on processed layers with the goal to show a remaining low damping for NiMnSb after processing. Therefore results of FMR measurements are presented which show besides a damping in the low 10^{-3} range the presence of higher modes. These modes are verified to be magnetostatic modes by performing computer based simulations using the program OOMMF (Object Oriented Micro Magnetic Framework) [Dona 99]. In the last part of the chapter transport measurements of fabricated spin-torque devices based on NiMnSb in combination with Py are presented. Magnetoresistance measurements show a high GMR ratio of 3,4 % at room temperature up to 7 % at low temperature. Current induced switching measurements show a low critical current necessary for switching the pseudo-spin-valve between the parallel and the anti parallel magnetic configuration. Besides DC measurements AC characteristics of the devices are shown as well. For such devices based on NiMnSb a q-factor of the emitted high frequency in the absence of an external field is measured to be an order higher than the highest published q-factor [Devo 07].

Chapter 2

Theory

Ferromagnetic resonance measurements are used in this thesis to determine the magnetic characteristics of NiMnSb. By doing an angle dependent measurement of the resonance field one can determine the anisotropies in the material. On the other hand by measuring the absorption line width for different frequencies the damping can be measured as well. In the following chapter the theoretical background is presented. It starts with energy considerations which are the origin for anisotropies in layers followed by mechanism which are the base for damping. Similar to the results chapter the theory of spin pumping and sinking is introduced shortly. As a last part of this theory chapter the giant magnetoresistance and spin torque theory are introduced.

2.1 Energy consideration

By performing an angle dependent scan of the resonance field in FMR measurements the curve of the resonance field as a function of angle gives the magnetic anisotropies in the ferromagnet and therefore offering a method for characterizing magnetostatic energy within the material. Subsequently the following section describes the magnetostatic energy in magnetic layers. Generally three different energy terms contributing to the total free magnetic energy ϵ_{tot}. In ferromagnetic materials the total energy is a sum of shape energy ϵ_{shape}, Zeeman energy ϵ_{zeeman} and crystal energy $\epsilon_{crystal}$. Whilst the last term is only present for ferromagnets with a crystal structure the first two terms are always applicable.

$$\epsilon_{tot} = \epsilon_{shape} + \epsilon_{zeeman} + \epsilon_{crystal} \tag{2.1}$$

2.1.1 shape anisotropy

Considering a uniform ferromagnetic layer which is sufficiently thin (the lateral dimension being much bigger than the thickness, $l_x \gg l_z \ll l_y$), the magnetization can be assumed to be in the plane. By bringing the ferromagnet into an external field with its direction out-of-plane of the layer, the magnetization will be tilted out of plane and magnetic charges on the surface of the slab will be generated. The resulting dipole field can be described:

$$\epsilon_{shape} = 2D\pi M_s^2 sin[\theta_M^2] = 2\pi M_\perp^2 \tag{2.2}$$

The parameter D is the effective demagnetization factor. Its value is very close to 1 for layers thicker than a few mono-layers. $M_\perp$ is the component of the magnetization perpendicular to the surface and θ_M is the angle between magnetization and surface, see fig. 2.1.

2.1.2 Zeeman energy

The Zeeman energy describes the energy splitting of a degenerated energy state by introducing an external field. This energy is described by the following equation:

$$\epsilon_{zeeman} = -\mathbf{H}_0 \cdot \mathbf{M} \tag{2.3}$$

2.1.3 crystal anisotropy

For ferromagnets with crystal structure energetic minimums and maximums are present due to symmetry considerations. Spin-orbit coupling is the reason for this anisotropy

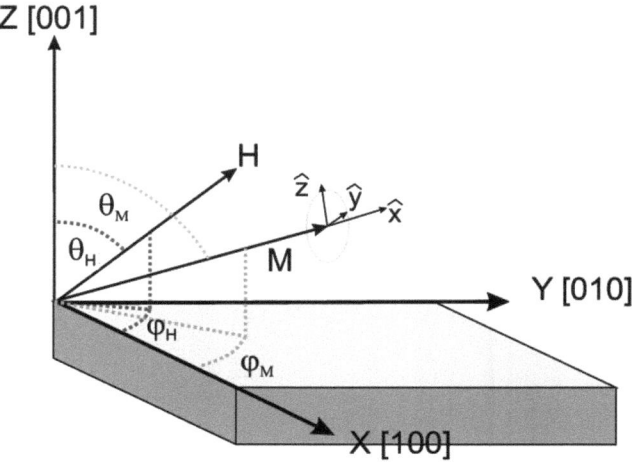

Fig. 2.1: Orientation of the magnetization and the external magnetic field

wherein the net spin moment is coupled to orbital motion given by the lattice potential. NiMnSb films exhibit a cubic anisotropy with a superimposed uniaxial anisotropy [Kove 05]. The crystal energy can be described using the following expression:

$$\epsilon_{crystal} = -\frac{K_1^{\parallel}}{2}\left(\alpha_x^4 + \alpha_y^4\right) - \frac{K_1^{\perp}}{2}\alpha_z^4 - K_u^{\perp}\alpha_z^2 \tag{2.4}$$

In equation 2.4, α_i, with $i = x, y, z$, are the cosines with respect to the crystal axis ([100],[010] and [001], respectively). The anisotropy constants $K_1^{\parallel}$, $K_1^{\perp}$ represent the fourfold in-plane and out-of-plane contribution (with respect to the (001) plane) and $K_u^{\perp}$ is the out-of-plane uniaxial anisotropy due to the thin films. For NiMnSb an additional uniaxial anisotropy in the plane can be described by adding another term, namely:

$$\text{uniaxial in-plane energy} = K_u^{\parallel}\frac{(\widehat{1}_{uni}M)^2}{M_s^2} \tag{2.5}$$

Hereby $\widehat{1}_{uni}$ is the unit vector along the in-plane uniaxial axis. For thin films one has to take into account that due to interface symmetry-breaking another anisotropy can come up. The strength of this constant is inverse proportional to the film thickness. Therefore the anisotropy constant for a thin films with thickness t and two interfaces I and II can be written in the form:

$$\kappa = K_{bulk} + \frac{K_I}{t} + \frac{K_{II}}{t} \tag{2.6}$$

2.1. Energy consideration

κ here stands for $K_1^{\parallel}$, $K_1^{\perp}$, $K_u^{\parallel}$ and $K_u^{\perp}$ whilst K_{bulk} and $K_{I,II}$ stand for the bulk and interface contribution, respectively.

Having discussed the magnetostatic energy considerations in the next section the dynamics in paramagnets and ferromagnets are presented.

2.2 Magnetization dynamics

If an external field $\mathbf{H}_0$ acts on a magnetic moment $\boldsymbol{\mu}$ the equilibrium state of the magnetic moment will be effected. The result is a torque between the external field and the magnetization. Classically a magnetic moment in an external magnetic field can be described using the Landau-Lifschitz equation [Land 35]:

$$\frac{\delta \boldsymbol{\mu}}{\delta t} = -\gamma \left(\boldsymbol{\mu} \times \mathbf{H}_0 \right) \tag{2.7}$$

with $\gamma = |g\mu_B/\hbar|$ the gyromagnetic ratio, g is the Landé g-factor, μ_B the Bohr magneton and $\hbar$ the Planck constant. This equation describes the motion of a magnetic moment without damping. For a system without damping the magnetization precesses with a constant angle between the direction of the magnetization and the applied field. By introducing damping the magnetization spirals down and aligns with the external field with a time dependence of $\tau = 1/G$, with G the Gilbert damping constant.

The coordinate system is defined in fig. 2.2. For NiMnSb it is sufficient to define the

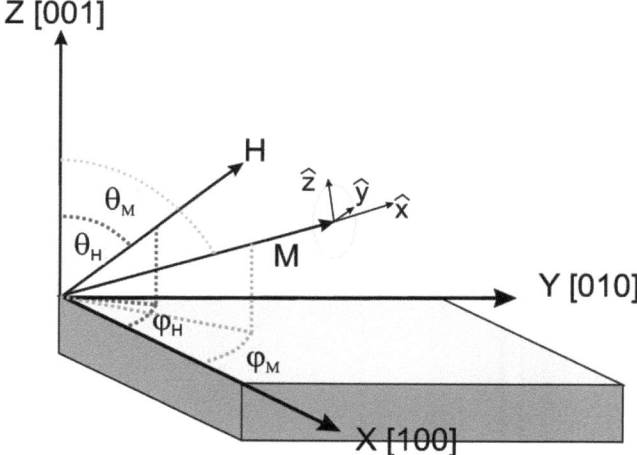

Fig. 2.2: Coordinate system which is used to describe motion of the magnetization and ferromagnetic resonance

angle with respect to a crystal axis. The global coordinate system is such that the axes (X,Y,Z) are aligned with the crystallographic axes [100], [010] and [001] respectively. The angles φ_i and θ_i with i=M,H are azimuthal and polar angle of the magnetization and the external magnetic field. The coordinate system $\widehat{x},\widehat{y}$ and $\widehat{z}$ is aligned with $\widehat{x}\|\mathbf{M}$ and the $\widehat{y}$ is parallel to the X-Y plane.

2.3. Ferromagnetic resonance (FMR)

The total magnetization of ferromagnets can be viewed as a superposition of individual magnetic moments.

$$\mathbf{M} = \sum \mu \tag{2.8}$$

The internal fields are introduced in the Landau-Lipschitz equation by exchanging the magnetic field $\mathbf{H}_0$ with an effective field $\mathbf{H}_{eff}$ which is defined as followed:

$$\mathbf{H}_{eff} = -\frac{\delta \epsilon_{tot}}{\delta \mathbf{M}} \tag{2.9}$$

With ϵ_{tot} defined in equation 2.1. Additionally the Landau-Lifschitz equation has to be extended by a damping term. The damping term can be written in the form:

$$T_D^{LL} = \lambda \left[\mathbf{H}_{eff} - \left(\mathbf{H}_{eff} \cdot \frac{\mathbf{M}}{M_s} \right) \frac{\mathbf{M}}{M_s} \right] \tag{2.10}$$

With λ as a phenomenological damping parameter. This damping term was introduced by Landau and Lipschitz and later modified by Gilbert to the well known Landau-Lifschitz-Gilbert equation. The second term on the right hand side can be rewritten in the form:

$$T_D^{LL} = -\frac{\lambda}{M_s^2} \mathbf{M} \times [\mathbf{M} \times \mathbf{H}_{eff}] \tag{2.11}$$

Gilbert formulated in 1955 an equivalent expression for the damping parameter

$$T_D^G = \frac{G}{M_s^2 \gamma} \mathbf{M} \times \frac{d\mathbf{M}}{dt} = \frac{\alpha}{M_s} \mathbf{M} \times \frac{d\mathbf{M}}{dt} \tag{2.12}$$

This expression uses the dimensionless damping parameter α. This parameter is measurable without needing to know the saturation magnetization and therefore accessible in our measurements.
By combining equation 2.7 and 2.12 one gets the commonly used Landau-Lifschitz-Gilbert (LLG) equation.

$$\frac{\delta \mathbf{M}}{\delta t} = -\gamma \left(\mathbf{M} \times \mathbf{H}_{eff} \right) + \frac{G}{\gamma M_s^2} \left(\mathbf{M} \times \frac{\delta \mathbf{M}}{\delta t} \right) \tag{2.13}$$

With the Gilbert damping constant G. The damping can equivalently expressed using the dimensionless damping parameter $\alpha = G/\gamma M_s$

2.3 Ferromagnetic resonance (FMR)

Ferromagnetic resonance can be measured using two different techniques: time-domain and frequency-domain. For time-domain measurements a rectangular pulse is applied to the sample which is saturated [1] by a static magnetic field. By measuring the precession of the magnetization the Larmor frequency can be extracted. The decay in time allows determination of the damping. In frequency-domain measurements the magnetic field is swept while a small high frequency field with fixed frequency is applied. By measuring the absorption of the alternating signal over the applied field the effective internal field can be measured. The line width of the absorption gives access to the damping.

2.3.1 resonance condition

In the following section the resonance condition is derived. The internal fields are described in Cartesian coordinates (see fig.2.2) and can be evaluated in terms of in-plane and out-of-plane angle as follows:

$$\alpha_x = \frac{M_x}{M_s} \cos\varphi_M \sin\theta_M - \frac{M_y}{M_s} \sin\varphi_M - \frac{M_z}{M_s} \cos\varphi_M \cos\theta_M \tag{2.14}$$

$$\alpha_y = \frac{M_x}{M_s} \sin\varphi_M \sin\theta_M + \frac{M_y}{M_s} \cos\varphi_M - \frac{M_z}{M_s} \sin\varphi_M \cos\theta_M \tag{2.15}$$

$$\alpha_z = \frac{M_x}{M_s} \cos\theta_M + \frac{M_z}{M_s} \sin\theta_M \tag{2.16}$$

The measurements presented in this thesis are all done in the in-plane configuration. For this geometry the out-of-plane angles $\theta_M = \theta_H$ are equal to 90° So the equations 2.14 to 2.16 for in-plane configuration are:

$$\alpha_x^{ip} = \frac{M_x}{M_s} \cos\varphi_M - \frac{M_y}{M_s} \sin\varphi_M \tag{2.17}$$

$$\alpha_y^{ip} = \frac{M_x}{M_s} \sin\varphi_M + \frac{M_y}{M_s} \cos\varphi_M \tag{2.18}$$

$$\alpha_z^{ip} = \frac{M_z}{M_s} \tag{2.19}$$

---

[1] in this thesis the term "saturated" describes the condition of all magnetic moments of a magnetic material being aligned to the external magnetic field

2.3. Ferromagnetic resonance (FMR)

With the aid of 2.4, 2.9 and 2.19 the anisotropy field can be written as:

$$H_x^{aniso} = \frac{\delta \epsilon_{aniso}}{\delta M_x} = \delta(-\frac{K_1^{\|}}{2}\left(\alpha_x^4 + \alpha_y^4\right) - \frac{K_1^{\perp}}{2}\alpha_z^4 - K_u^{\perp}\alpha_z^2)/\delta M_x$$

$$= \frac{K_1^{\|}}{2M_s^4}\left(M_x^3(\cos 4\varphi_M + 3) - 3M_x^2 M_y \sin 4\varphi_M\right.$$
$$\left. + M_x M_y^2(3 - 3\cos 4\varphi_M) + M_y^3 \sin 4\varphi_M\right)) \qquad (2.20)$$

$$H_y^{aniso} = \frac{\delta \epsilon_{aniso}}{\delta M_y} = \delta(-\frac{K_1^{\|}}{2}\left(\alpha_x^4 + \alpha_y^4\right) - \frac{K_1^{\perp}}{2}\alpha_z^4 - K_u^{\perp}\alpha_z^2)/\delta M_y$$

$$= \frac{-K_1^{\|}}{2M_s^4}\left(M_x^3 \sin 4\varphi_M - M_x^2 M_y(3 - 3\cos 4\varphi_M)\right.$$
$$\left. - M_x M_y^2(3 + 3\sin 4\varphi_M) - M_y^3(\cos 4\varphi_M + 3)\right)) \qquad (2.21)$$

$$H_z^{aniso} = -4\pi D M_z + \frac{M_u^{\perp}}{M_s^2} M_z + \frac{2K_{1\perp}}{M_s^4} M_z^3 \qquad (2.22)$$

The external field can be expressed in terms of φ_M and φ_H.

$$\mathbf{H}_0 = H_0(\cos(\varphi_M - \varphi_H)\hat{x} - \sin(\varphi_M - \varphi_H)\hat{y}) \qquad (2.23)$$

The effective field $\mathbf{H}_{eff}$ can be described as the sum of all acting fields. The small RF-field $\mathbf{h}(t)$ which is in the film plane also needs to be taken into account.

$$\mathbf{H}_{eff} = \mathbf{H}_0 + \mathbf{H}^{aniso} + h\hat{y} \qquad (2.24)$$

To calculate the resonance condition a small angle approximation ($M_x \gg M_y$, M_z) is used for an applied field along the x-axis as shown in fig.2.2.

$$\mathbf{M} = M_s \hat{x} + m_y^{rf} \hat{y} + m_z^{rf} \hat{z} \qquad (2.25)$$

After inserting equ. 2.24 and 2.25 in the LLG equation one gets a system of coupled equations for the magnetization. The solution of this system can be linearised because the components $m_{x,y}^{rf}$ are small compared to the saturation magnetization.

$$0 = \frac{\omega}{\gamma} m_y^{rf} + \left(\mathcal{B}_{eff} + i\alpha\frac{\omega}{\gamma}\right) m_z^{rf} \qquad (2.26)$$

$$hM_s = -\frac{\omega}{\gamma} m_z^{rf} + \left(\mathcal{H}_{eff} + i\alpha\frac{\omega}{\gamma}\right) m_y^{rf} \qquad (2.27)$$

here the effective b-field $\mathcal{B}_{eff}$ and the effective h-field $\mathcal{H}_{eff}$ are defined as follows.

$$\mathcal{B}_{eff} = H_0 \cos(\varphi_M - \varphi_H) + 4\pi D M_s - \frac{2K_u^{\perp}}{M_s}$$
$$+ \frac{K_1^{\|}}{2M_s}[3 + \cos 4\varphi_M] + \frac{K_u^{\|}}{M_s}[1 + \cos 2(\varphi_M - \varphi_U)] \qquad (2.28)$$

$$\mathcal{H}_{eff} = H_0 \cos(\varphi_M - \varphi_H) + \frac{2K_1^{\|}}{M_s}\cos 4\varphi_M + \frac{2K_u^{\|}}{M_s}\cos 2(\varphi_M - \varphi_U) \qquad (2.29)$$

Because the saturation magnetization $4\pi DM_s$ and the out-of-plane uniaxial anisotropy contribute additively to the effective B-field it is common to define an effective magnetization $4\pi M_{eff}$ as follows:

$$4\pi M_{eff} = 4\pi DM_s - \frac{2K_u^\perp}{M_s} \qquad (2.30)$$

With D as the demagnetization factor. For a given frequency ω the RF-susceptibility can be calculated to be:

$$\chi_y = \frac{m_y^{rf}}{h} = \frac{M_s\left(\mathcal{B}_{eff} - i\alpha\frac{\omega}{\gamma}\right)}{\left(\mathcal{B}_{eff} - i\alpha\frac{\omega}{\gamma}\right)\left(\mathcal{H}_{eff} - i\alpha\frac{\omega}{\gamma}\right) - \left(\frac{\omega}{\gamma}\right)^2} \qquad (2.31)$$

Equation 2.31 is only valid for a uniform RF-field. This is only the case for very thin layers in which the skin depth is much bigger than the layer thickness ($t_{skin} \gg t_{layer}$). Nevertheless the expression given above holds for the layers measured in this thesis. The absorption has a divergence if the denominator is zero therefore the resonance condition is as follows (setting α=0, no damping):

$$\left(\frac{\omega}{\gamma}\right)^2 = \mathcal{B}_{eff}\mathcal{H}_{eff} \qquad (2.32)$$

The measurements in this thesis are all carried out at frequencies between 8 and 15 GHz (which are low compared to standard frequencies of up to 72 GHz). Because of such low frequencies and consequently the corresponding small field necessary to achieve the resonance conditions is small as well, the external field can not be assumed to be parallel to the magnetization of the sample, $\varphi_M \neq \varphi_H$. The angle φ_M can be found by assuming the absence of an acting torque on the magnetization in the static equilibrium (M_x=M_s,M_y=0,M_z=0). After inserting the effective fields (eqs. 2.20 to 2.22 and 2.23) in equation 2.13 and setting the z-component to zero:

$$H_0 M_s \sin(\varphi_H - \varphi_M) + K_u^{\|}\sin 2(\varphi_U - \varphi_M) - \frac{1}{2}K_1^{\|}\sin 4\varphi_M = 0 \qquad (2.33)$$

The imaginary part of the microwave susceptibility is an almost perfect Lorentzian for the saturated case ($\mathbf{M} \parallel \mathbf{H}_0$) and can be described as:

$$Im[\chi_y] = M_s \left.\frac{\mathcal{B}_{eff}}{\mathcal{B}_{eff} + \mathcal{H}_{eff}}\right|_{H_{FMR}} \frac{\Delta H}{\Delta H^2 + (H_0 - H_{FMR})^2} \qquad (2.34)$$

Here $\Delta H = \alpha\frac{\omega}{\gamma}$ is the half width half maximum (HWHM) and H_{FMR} is the peak position.

2.3.2 damping mechanisms

The intrinsic damping can be divided in three major contributing mechanisms (i) eddy currents, (ii) magnon-phonon scattering and (iii) electron relaxation processes.

2.3. Ferromagnetic resonance (FMR)

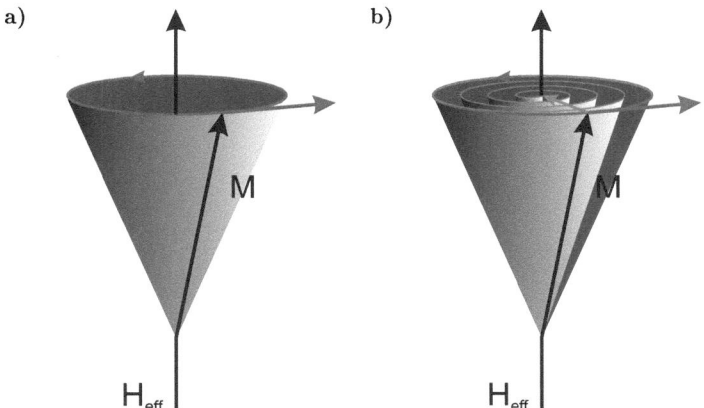

Fig. 2.3: a) magnetization motion without damping; b) motion of the magnetization with damping, blue arrow indicating the damping torque which is oriented onto the direction of the effective field

i eddy currents
An applied alternating field in the GHz range causes eddy currents. These eddy currents can be expressed as an extended damping mechanism and manifest itself as a broadening of the line width. The eddy current dependent part can be written as:

$$\alpha_{eddy} = \frac{1}{6}\left(\frac{4\pi}{c}\right)^2 \sigma d^2 \qquad (2.35)$$

Here is σ the electrical conductivity and d the thickness of the film.

ii magnon-phonon scattering
Another contribution to the damping is the magnon-phonon scattering. This can be described using:

$$\alpha_{magnon} = \frac{2\eta\gamma}{M_s}\left(\frac{B_2^2(1+\nu)^2}{E^2}\right) \qquad (2.36)$$

With η is the magnon phonon viscosity and B_2 is the magnetoelastic shear constant.

iii electron relaxation processes
Besides the mentioned mechanism for damping in thin ferromagnetic layers the most important mechanism is based on electron mechanisms, namely intraband and

interband transitions. Bret Heinrich [Hein 67] proposed a model for such transition based on interaction with s-p like electrons with localized spins. This results to additional Gilbert damping parameters α for intra and interband transitions

$$\alpha_{SO}^{intra} \simeq \frac{\langle S \rangle^2}{M_s \gamma} \left(\frac{\xi}{\hbar}\right)^2 \left(\sum_{\mu} \chi_P^{\mu} \langle \mu | L^+ | \mu \rangle \langle \mu | L^- | \mu \rangle \right) \tau_m \qquad (2.37)$$

and

$$\alpha_{SO}^{inter} \simeq \frac{\langle S \rangle^2}{M_s \gamma} (\Delta g_\alpha)^2 \frac{1}{\tau_m} \qquad (2.38)$$

Hereby χ_P^{μ} is the Pauli susceptibility of the given Fermi sheet and τ_m is the relaxation time. ξ is the coefficient of spin-orbit interaction and $L^{\pm}$ is the left and right handed component of the atomic site transverse angular momentum.

2.4 Concept of spin pumping and spin sinking

In 2001 it was experimentally observed by Mizukami and co-workers that a ferromagnet in contact to a non-magnetic metal exhibits a different damping constant (the group measured thickness dependent linewidth of the ferromagnetic resonance for Permalloy gapped with different non-magnetic materials and found an increased linewidth with decreased layer thickness which was also connected to the gapped material [Mizu 01b, Mizu 01a]). The damping constant of the heterostructure is, for thin ferromagnets (< 10 nm), larger then the damping measured for bulk ferromagnets. Tserkovnyak et al. [Tser 02] evaluated a model describing the reason for this additional damping. The concept is reversal of the concept of switching a ferromagnet by passing a spin current. Like the spin current can transfer momentum and therefore switch a ferromagnet a precessing ferromagnet can pump spin current in a metal which is in contact to it.

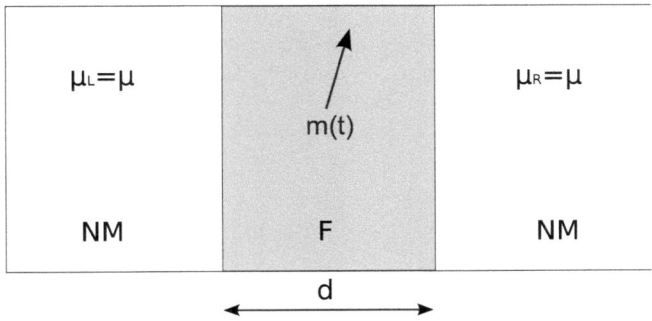

Fig. 2.4: A ferromagnetic layer (F) sandwiched between two paramagnet electron reservoirs (NM) with equal energy (being in thermal equilibrium)

The fig. 2.4 shows a ferromagnetic layer which can pump spin current in the electron reservoirs attached to it. A 2×2 current matrix in spin space can be written as follows:

$$\widehat{I} = \widehat{1}\frac{I_c}{2} - \widehat{\sigma}\frac{I_s e}{\hbar} \qquad (2.39)$$

With I_c for the charge current and I_s for the spin currents, $\widehat{1}$ is the unit matrix and $\widehat{\sigma}$ is the vector of Pauli spin matrices. For the absence of an applied voltage and a change in the external field the charge current vanishes, I_c=0. For the spin current I_s one has to distinguish two contributions:

- (1) I_s^{pumped} which is the current pumped into the normal metal and

- (2) I_s^0 which is the current flowing back in the ferromagnet.

Using an adiabatic approximation the time dependent current I(t) pumped by the precession of the magnetization in the ferromagnet can be calculated since the period of precession $2\pi/\omega$ is larger than the relaxation time.

$$I_s^{pumped}(t) = \frac{\hbar}{4\pi}\left(A_r m \times \frac{dm}{dt} - A_i \frac{dm}{dt}\right) \quad (2.40)$$

Hereby A_r and A_i are the interface parameters:

$$A_r = \frac{1}{2}\sum_{mn}\{|r_{mn}^\uparrow - r_{mn}^\downarrow|^2 + |t_{mn}^\uparrow - t_{mn}^\downarrow|^2\} \quad (2.41)$$

$$A_i = Im\sum_{mn}\{r_{mn}^\uparrow(r_{mn}^\downarrow)^\star + t_{mn}^\uparrow(t_{mn}^\downarrow)^\star\} \quad (2.42)$$

With $r_{mn}^\uparrow$ ($r_{mn}^\downarrow$) is the reflection coefficient for spin-up (spin-down) and $t_{mn}^\uparrow$ ($t_{mn}^\downarrow$) the transmission coefficient for electrons in the left (or right) lead. For steady state state, $dm/dt=0$, the spin current according to equation 2.40 vanishes. For a precessing system which losses spin momentum into the normal metal leads conservation of angular momentum leads to an additional term in the LLG descries the spin torque on the ferromagnet. This additional term effects the gyromagnetic ratio and the damping as follows:

$$\frac{1}{\gamma} = \frac{1}{\gamma_0}\left\{1 + g\frac{A_i^{(L)} + A_i^{(R)}}{4\pi M}\right\} \quad (2.43)$$

$$\alpha = \frac{\gamma}{\gamma_0}\left\{\alpha_0 + g\frac{A_r^{(L)} + A_r^{(R)}}{4\pi M}\right\} \quad (2.44)$$

With g is the LandÃľ factor, M is the saturation magnetization in units of μ_B, γ_0 and α_0 are the gyromagnetic ratio and the damping constants for the bulk material. It can be seen in equation 2.43 and 2.44 that $A_i^{(L)}+A_i^{(R)}$ effecting the position by an additive term to $1/\gamma$ whilst $A_r^{(L)}+A_r^{(R)}$ effects the line width.

2.5 Giant magnetoresistance (GMR)

The GMR effect has its origin in two separate physical phenomena. A description of the GMR effect can be made on the basis of the bandstructure (density of states, Fermi velocity, quantum interference) or on scattering processes. The first effect is called *intrinsic* and the second one is called *extrinsic* GMR.

2.5.1 Mott two spin channel-model

One model is the Mott Model of two spin-channels to describe the GMR effect. The model is based on different scattering probabilities (τ) for spins with different momentum on a magnetic layer with various magnetization. Taking τ_{even} as the scattering rate for an electron with spin direction parallel to the magnetization of the layer and τ_{odd} for an electron with antiparallel alignment of its spin. The scattering rate for electrons with antiparallel alignment is assumed to be bigger than for parallel alignment and therefore the specific resistance ρ_{even} smaller than the specific resistance ρ_{odd}. The schematic for an equivalent ohmic circuit is shown on fig.2.5a and fig.2.5b, respectively.

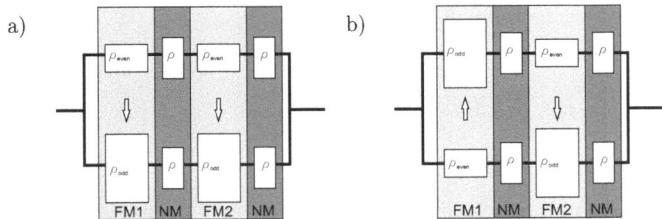

Fig. 2.5: schematic giant magnetoresistance for a) parallel and b) antiparallel alignment

According to the schematic one can calculate the resistance for the parallel state to be:

$$\rho_p = \left(\frac{1}{2\rho_{even}} + \frac{1}{2\rho_{odd}} \right)^{-1} = \frac{2\rho_{even}\rho_{odd}}{\rho_{even} + \rho_{odd}} \qquad (2.45)$$

For the antiparallel alignment the total resistance can be calculated to:

$$\rho_{ap} = \left(\frac{1}{\rho_{even} + \rho_{odd}} + \frac{1}{\rho_{odd} + \rho_{even}} \right)^{-1} = \frac{\rho_{even} + \rho_{odd}}{2} \qquad (2.46)$$

The GMR ratio can be defined as followed:

$$GMR \equiv -\frac{\rho_p - \rho_{ap}}{\rho_p} = \frac{\rho_{ap} - \rho_p}{\rho_p} = \frac{\sigma_p}{\sigma_{ap}} - 1 \qquad (2.47)$$

Using equation 2.45 and 2.46 one can rewrite equation 2.47 in the form:

$$GMR = \frac{(\rho_{even} - \rho_{odd})^2}{4\rho_{even}\rho_{odd}} \qquad (2.48)$$

Taking into account the specific resistance of the non-magnetic materials the resistance of the parallel state can be calculated to be:

$$R_p = \frac{(1+\gamma_{GMR}a)(1+\gamma_{GMR}b)}{2+\gamma_{GMR}a+\gamma_{GMR}b}\frac{2\rho d}{A} \qquad (2.49)$$

and the resistance for the antiparallel state to be:

$$R_{ap} = (2 + \gamma a + \gamma b)\frac{\rho d}{A} \qquad (2.50)$$

Here is $\gamma_{GMR} = d/d_M$ with d and d_M the thickness of the non-magnetic and the magnetic layers respectively, and $a = \rho_{even}/\rho$ and $b = \rho_{odd}/\rho$ and A the cross-section area. Using equation 2.49 and 2.50 the GMR ratio can be expressed, including the resistance of the non-magnetic layers, as follows:

$$GMR = \frac{R_{ap} - R_p}{R_p} = \frac{(a-b)^2}{4(a+\gamma_{GMR})(b+\gamma_{GMR})} \qquad (2.51)$$

2.5.2 intrinsic GMR

As mentioned above the intrinsic giant magnetoresistance has its origin in the bandstructure of the heterostructure.

By bringing a ferromagnet in an external field the energy of the spins aligned with the field is lowered (majority spins, indicated by ↑) whilst the energy for the spins antiparallel to the field is increased (minority spins, indicated by ↓). The difference in the energy is equalized by spin flip processes. In fig. 2.6 a bandstructure is shown in which the splitting, induced by an external field H with an amount of E=μ_BH, is shown. The hatched area indicates the difference and the spins which had flipped its orientation. Assuming the two-channel spin model the conductivities for majority and minority spin channels sum up to the total conductivity.

$$\sigma_{tot} = \sigma_\uparrow + \sigma_\downarrow \qquad (2.52)$$

Using the Boltzmann-transport theory the conductivity can be calculated.

$$\hat{\sigma} = \frac{2e^2\tau}{V}\sum_k \mathbf{v}(\mathbf{k})\mathbf{v}(\mathbf{k})\delta(E(\mathbf{k}) - E_F) \qquad (2.53)$$

2.5. Giant magnetoresistance (GMR)

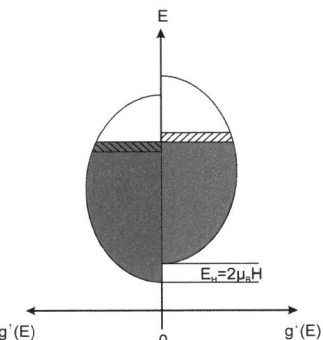

Fig. 2.6: Band structure of a ferromagnet in an external field, the external field is parallel to the spin ↑ and therefore the energy is lowered by an amount

The components of the conduction tensor $\hat{\sigma}$ are different for transport perpendicular to plane and for in-plane.

$$\hat{\sigma} = \begin{pmatrix} \sigma_{xx}^{cip} & 0 & 0 \\ 0 & \sigma_{yy}^{cip} & 0 \\ 0 & 0 & \sigma_{zz}^{cpp} \end{pmatrix} \tag{2.54}$$

Therefore the GMR ratio can be written:

$$GMR = \frac{\sum_k \delta(E^\uparrow(k) - E_F)(v_{k_i}^\uparrow)^2 + \sum_k \delta(E^\downarrow(k) - E_F)(v_{k_i}^\downarrow)^2}{2\sum_k \delta(E^{ap}(k) - E_F)(v_{ki}^{ap})^2} - 1 \tag{2.55}$$

The v_{k_i} are the Cartesian components of the Fermi velocity. Using $N^{\uparrow/\downarrow}(E_F) = \sum_k \delta(E^{\uparrow/\downarrow}(k) - E_F)$ and the averaged squares over the Fermi surface of the components of the velocity $\left\langle v_{ki}^{\uparrow/\downarrow 2} \right\rangle$ we get for the GMR ratio:

$$GMR = \frac{N^\uparrow(E_F)\left\langle v_{ki}^{\uparrow 2} \right\rangle + N^\downarrow(E_F)\left\langle v_{ki}^{\downarrow 2} \right\rangle}{2N^{ap}(E_F)\left\langle v_{ki}^{ap2} \right\rangle} - 1 \tag{2.56}$$

This equation shows that the GMR ratio is directly connected to the velocities of the majority and minority electrons at the Fermi-edge and their density of states.

2.5.3 extrinsic GMR

The extrinsic giant magnetoresistance is a result of the scattering of electron waves at impurities.

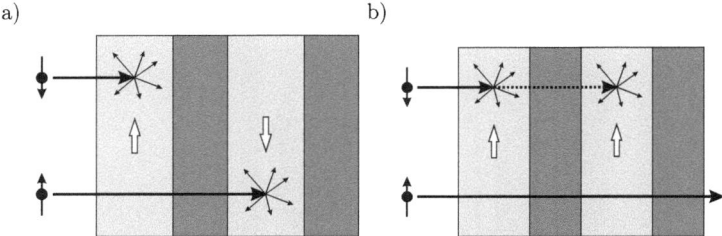

Fig. 2.7: Scattering for antiparallel alignment (a) and for parallel alignment (b)

There are two main reasons why the majority spins are less scattered then the minority spins:

- the states are mostly occupied for spin obtaining scattering and
- the spin flip scattering is low (see Annex A.1)

The total conductance for the parallel case is dominated by the majority spins $\sigma^p = \sigma^{p\uparrow} + \sigma^{p\downarrow}$ with $\sigma^{p\uparrow} > \sigma^{p\downarrow}$. For the antiparallel alignment the conductivity for either the majority or the minority channel is equal to the total conductivity $\sigma^{ap\uparrow} = \sigma^{ap\downarrow} = \sigma^{ap}$. The total giant magnetoresistance effect is based on both mechanisms, the intrinsic and the extrinsic effect.

2.6 Spin-torque

In 1996 Slonczewski [Slon 96] and Berger [Berg 96] independently realized that a current which passes a perpendicular to plane spin valve produces a spin torque on the magnetic moments of the ferromagnets and that this torque can switch the magnetizations or drive high frequency oscillations. Since the current densities which are necessary to change the magnetic formation were calculated to be in the order of $10^8 A/m^2$ the experimental realization followed shortly after. The first observations were done at nano-point contacts by Tsoi et al in 1998 [Tsoi 98]. Afterwards in the early 1999 Sun and co-workers [Sun 99] presented current induced switching in tunnel magneto resistance (TMR) structures with dimensions in the μm^2 range. After the first experiments using TMR structures Myers et al [Myer 99] demonstrated current induced switching in GMR devices the first time in nano-point contact geometry. Shortly followed by the first current assistant switching experiments in nanopillar geometry using Co/Cu/Co multilayers and a diameter of 100 nm was published by Katine et al [Kati 00].

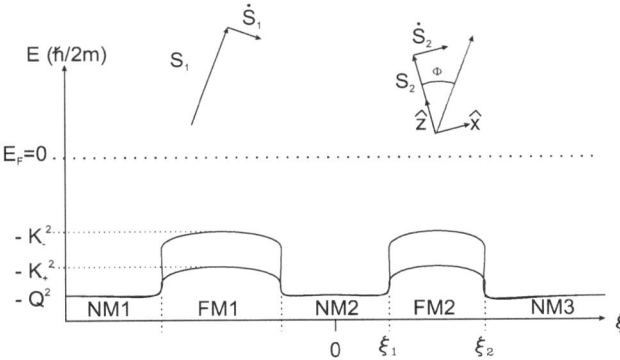

Fig. 2.8: Coulomb and exchange potentials versus position ξ in a five layer system is shown whilst the particle flow is rightward the charged flow is leftward. Vector diagram shows the spin moments $S_{1,2}$ and their current driven velocities $\dot{S}_{1,2}$ (from [Slon 96])

Considering five regions consisting of non-magnetic layers (NM1, NM2 and NM3) and two ferromagnetic layers (FM1 and FM2) as shown in fig.2.8. ξ is the direction perpendicular to the planes of the multilayers. The magnetization vectors for the ferromagnetic layer FM1 and FM2 form the angle θ. If an electron is injected from the left side and the thickness of the non-magnetic layer NM2 between the two ferromagnets is less than the spin diffusion length of the material (for Cu $\lambda_s \approx 1\mu m$ at room temperature [Jede 01]) the polarization of the electron (received from the first ferromagnetic layer FM1) partly remains at FM2. To be more precise, a fraction of electrons injected in FM2 will have the orientation parallel to $\mathbf{S}_1$. A rotating coordinate system with $\widehat{x}, \widehat{y}, \widehat{z}$ as unit vectors is used. The unit vector $\hat{z}$ is parallel to the magnetic moment of the second ferromagnet FM2 $\mathbf{S}_2 = S_2 \hat{z}$. The direction $\hat{y}$ is defined by the vector product $\mathbf{S}_1 \times \mathbf{S}_2$. Using $(cos(\theta), sin(\theta))$ for the spin orientation of the incident electrons. One can calculate the wave numbers $k_\pm(\xi)$ by using the following equation:

$$k_{\uparrow/\downarrow}(\xi) = (E - k_p^2 - V_{\uparrow/\downarrow})^{1/2} \cdot \frac{(2m_e)^{1/2}}{\hbar} \tag{2.57}$$

Here $V_{\uparrow/\downarrow}$ are the Coulomb plus Stoner exchange potential diagonals and the subscript $\uparrow/\downarrow$ representing the majority and minority spins. Starting with a three layer geometry (NM2, FM2 and NM3) with a given $\xi_0 = 0$ and the magnetic region between ξ_1 and ξ_2 (see fig. 2.8) and $V_\uparrow = V_\downarrow$ and $k_\uparrow = k_\downarrow$. The Hartree-Fock spinor wave function is:

$$\psi(\xi) = \begin{pmatrix} k_\uparrow^{-1/2}(\xi) exp\left(i \int_0^\xi d(\xi') k_\uparrow(\xi')\right) cos(\theta/2) \\ k_\downarrow^{-1/2}(\xi) exp\left(i \int_0^\xi d(\xi') k_\downarrow(\xi')\right) sin(\theta/2) \end{pmatrix} \tag{2.58}$$

For the particle flux one can write

$$\Phi_{\rightarrow,z}(\xi) = Im\left(\psi_\uparrow^* \frac{d\psi_\uparrow}{d\xi} \pm \psi_\downarrow^* \frac{d\psi_\downarrow}{d\xi}\right) \tag{2.59}$$

$$\Phi_\uparrow(\xi) = \Phi_\uparrow + i\Phi_y = i\left(\frac{d\psi_\uparrow^*}{d\xi}\psi_\downarrow - \psi_\uparrow^* \frac{d\psi_\downarrow}{d\xi}\right) \tag{2.60}$$

The Pauli-Spin flux within the regions of the non-magnetic layers (NM2 and NM3) is:

$$\Phi_\uparrow = exp\left(i \int_0^\xi (k_\downarrow - k_\uparrow) d\xi\right) sin(\theta) \tag{2.61}$$

$$\Phi_z = cos(\theta) \tag{2.62}$$

The equation 2.61 describes a precession cone of the magnetic moment of the spin about $\mathbf{S}_2$. By conservation of angular moment this requires a reaction of the $\mathbf{S}_2$ of $\Delta \mathbf{S}_2$ equal to the sum of the flux incoming on the left side and the right side:

$$\Delta \mathbf{S}_{2,x} + i\Delta \mathbf{S}_{2,y} = [\Phi_\uparrow(0) - \Phi_\uparrow(\infty)]$$

$$= \frac{1}{2}\left(1 - exp\left(i \int_0^\infty (k_\downarrow - k_\uparrow) d\xi\right)\right) sin(\theta) \tag{2.63}$$

$$\Delta \mathbf{S}_{2,z} = 0 \tag{2.64}$$

2.6. Spin-torque

Considering a ferromagnet with 100% spin polarization, which means $V_\downarrow > E$, than $k_\downarrow$ would be imaginary according to equation 2.57. Assuming the thickness of the ferromagnet FM2 is sufficiently thick to avoid tunneling processes of the minority spins, the wave function $\psi_\downarrow$ will be totally reflected while $\psi_\uparrow$ will be totally transmitted. Therefore the spin factor of $\psi_\downarrow$ is $(0, sin(\theta))$ and the factor for $\psi_\uparrow$ is $(cos(\theta), 0)$. Equation 2.63 becomes:

$$\Delta \mathbf{S}_2 = \frac{sin(\theta)}{2cos^2(\theta/2)}(1, 0, 0) \tag{2.65}$$

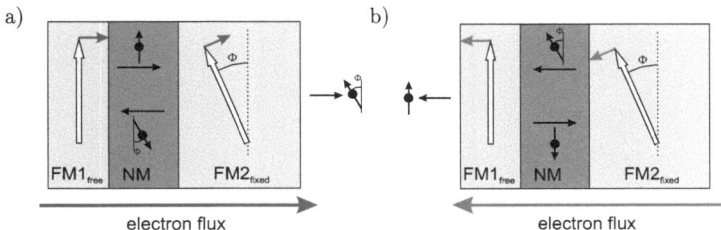

Fig. 2.9: An electron flux from the free to the fixed layer results in a torque on the free layer which points away from the fixed layer resulting in an antiparallel alignment (a) while an electron flux in the opposite direction results in a favoured parallel alignment due to direction of the resulting torque (b)

Equation 2.65 and equ.2.63 describe the total transmission and reflection of an electron flux on a ferromagnetic interlayer without polarization of the flux. By introducing a polariser as indicated in fig.2.8 a defined motion of the second ferromagnet can be realized.

Chapter 3

Samples

The following section introduces the NiMnSb half-Heusler alloy. Its crystal structure is displayed and its growing is described in a short overview. For a more profound dealing with the growth of NiMnSb the reader is guided to the dissertation thesis of Florian Lochner who grew the high quality NiMnSb layers which are measured in this thesis [Loch 10b]. Afterwards different types of spin-valves are presented and typical measurement curves (magnetization vs. B-field and resistant vs. B-field) are shown.

3.1 The half-Heusler alloy NiMnSb

NiMnSb belongs to the class of so called half-Heusler alloy [Heus 03]. Calculations used the augmented-spherical-wave method of Williams, Kübler and Gelatt done by de Groot et al. [DeGr 83] have predicted the half-metallic character of NiMnSb. This half-metal character makes it a very interesting material for spintronic devices. 2001 Wim van Roy et al. grew NiMnSb [Van 00, Van 01] on (001) GaAs substrates. A major disadvantage of using GaAs substrates was the big lattice mismatch of 4.4 %.

Spin-polarized inverse photoemission (SPIPES) experiments done by Borca and co-workers indicated an additional phase transition under well under the curie temperature and therefore the authors claimed the half-Heusler compound NiMnSb not being half-metallic at room temperature [Borc 01]. This non-half-metallic behaviour was based on various measurements of the polarization such as using Andrev reflection resulting in a polarization of only 58±2.3% [Soul 98] and to be 50% [Bona 85] from spin-polarized photoemission. This is consistent with low GMR ratio measured for perpendicular to plane transport through spin valves with NiMnSb as polariser [Caba 98]. Besides the low GMR ratio at room temperature low TMR ratio indication the lack of full spin polarization as well [Tana 99]. NiMnSb layers measured in this thesis were grown on an InP substrates with an (In,Ga)As buffer. This reduces the lattice mismatch and ensures a higher crystalline quality (lattice constant of InP 5.868 Å and of NiMnSb 5.906 Å).

Because of the theory predicts 100 % spin polarization for half-metals grown on a semiconductor only in (111) crystal direction in the year 2002 Wim van Roy and co-workers started to grow NiMnSb on (111)GaAs [Van 02, Van 03]. In this thesis measurements of (111)NiMnSb layers grown on an InP substrate with a 2% miscut in ($1\bar{1}2$) direction are presented as well. The miscut in the substrate ensures a good growth start and is responsible for a uniaxial anisotropy.

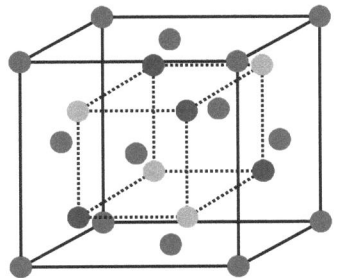

Fig. 3.1: C1$_b$ structure of NiMnSb (red:Ni, blue:Mn and green:Sb)

3.1.1 crystal structure of NiMnSb

NiMnSb is growing in C1$_b$ structure [Heus 03, Otto 89] In fig. 3.1 the crystal structure of NiMnSb is shown (red:Ni(0,0,0),blue:Mn(1/4,1/4,1/4) and green:Sb(3/4,3/4,3/4)).

3.1.2 growth of NiMnSb

Only a short summary of the growth of NiMnSb is presented here. The more interested reader is guided to the PhD-thesis of Florian Lochner [Loch 10b].

In the first step the oxide is removed from the InP wafer by heating the substrate to 520 °C. Afterwards the (In.Ga)As buffer layer is grown at a temperature range between 510 °C and 530 °C. The final step is to grow the NiMnSb layer at a temperature of 250 °C with flux ratios of Sb:Ni = 14.5:1 and Mn:Ni = 4.8:1. According to the presented FMR measurements done on these layers the crystal quality of NiMnSb grown on (001) (In,Ga)As is highly crystalline and offering a very low damping.

3.2 Different types of spin-valves

By using two ferromagnets, one acting as polariser and the second as analyser, in combination of a metallic interlayer a spin-valve can be realized. The most distinctive attribute of such a device are the two resistance states for the ferromagnets being parallel (low resistance) or anti-parallel (high resistance) aligned. In common devices in the industry like memory devices [Tehr 99, Zhu 00] mostly the spin-valve structure [Nogu 99] is used, but pseudo-spin-valves can be used as well [Ever 98]. The presented devices in this thesis are pseudo-spin-valves. Therefore the following section describes the differences between spin-valve and pseudo-spin-valve structure. A major difference between the spin-valves and the pseudo-spin-valves is an additional antiferromagnet for pinning the fixed electrode. In the pseudo-spin-valves a pinning is done either by choosing a hard (high coersive field) ferromagnet for the fixed layer or by changing the anisotropy in the layer, for example by the thickness. Nevertheless in a pseudo-spin-valve the fixed layer is not a 'real' fixed layer but 'more fixed' than the free layer. Consequently it is possible to switch the polariser and the analyser. To demonstrate the different behaviour magnetoresistance and magnetization curves for spin-valves and pseudo-spin-valves are shown.

3.2.1 Spin-valves

Fig.3.2 show the magnetoresistance and magnetization curves for a spin-valve.

In section I both layers are aligned in parallel and therefore the resistance of the device is in the low state (see chapter 2.5). In section II the analyser, the free layer respectively, is switched forming an anti-parallel alignment with the high resistance state. For fields high enough to overcome the interlayer coupling between the antiferromagnet and the 'fixed' ferromagnetic layer the device is in a parallel formation again reaching a low resistance state. The magnetization curve (fig. 3.2b)) shows the total magnetization of the device. The value for the saturation magnetization in the high negative field is equal to the value of the saturation for high positive fields.

Fig. 3.3 shows the minor loop and the complete loop for the hysteresis of a spin-valve structure. In fig.3.3.a the external magnetic field is swept back before aligning the antiferromagnet whilst in fig.3.3.b the external field is further increased until the exchange field of the antiferromagnet is overcame and the antiferromagnet is aligned parallel to the external field.

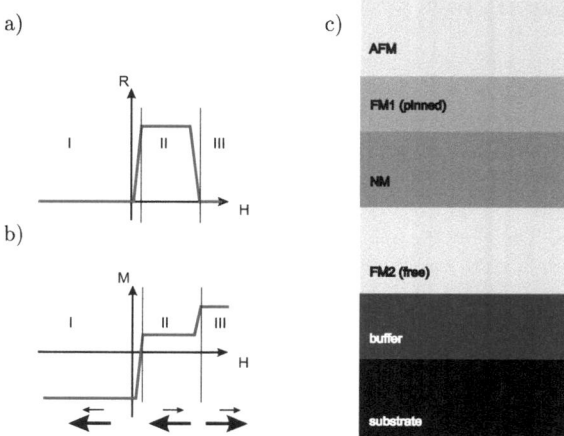

Fig. 3.2: a) Magnetoresistance and b) magnetization curve for a spin-valve structure with antiferromagnet (AFM) to 'pin' the fixed layer (the curves show field dependent measurement in which the field is lower than the exchange field of the antiferromagnet), in c) a typical heterostructure is shown

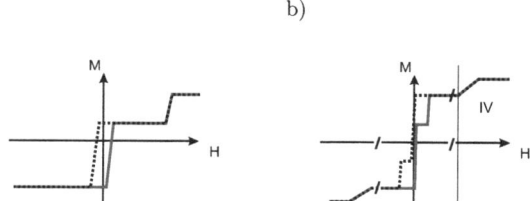

Fig. 3.3: a) magnetization minor loop for spin valve with antiferromagnet and b) complete loop hysteresis; in region IV the field is strong enough to align the antiferromagnet and therefore adding an extra magnetic moment to the magnetization

3.2.2 Pseudo-spin-valve

In contrast to the spin-valve, in which an antiferromagnet is exchange coupled to the 'fixed' ferromagnet, can a distinguishable switching be achieved in pseudo-spin-valves by choosing two ferromagnets with different coercive field or by defining structures with a shape anisotropy. This results in a much smaller plateau in the magnetoresistance plot.

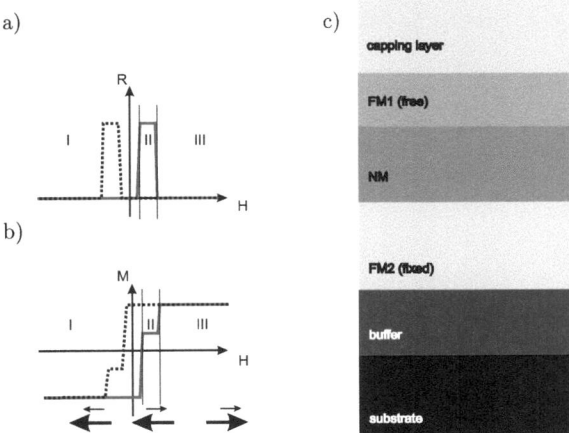

Fig. 3.4: magnetoresistance and magnetization curve for a pseudo-spin-valve structure. The difference in the switching fields can be achieved by either ferromagnet with different coercive fields or by engineering an additional shape anisotropy

Chapter 4

Results

In the following chapter the results are presented. After an introduction of the setup which was used to measure the ferromagnetic resonance the results of FMR measurements of (001) and (111)NiMnSb layers are presented followed by measurements of sub-micrometer sized elements. These FMR measurements indicate that the NiMnSb damping parameter is even lower than MBE grown iron (see. 4.2.1). As a direct result of the low damping magnetostatic modes could be measured in sub-micrometer sized elements. The main part of the chapter are the measurements of NiMnSb based spin-torque devices. Both the DC-experiments and the AC-measurements indicate the potential of NiMnSb as a novel material in spintronic devices.

Besides the results presented in this thesis, vortex experiments have been done at the Commissariat á l'Énergie Atomique (CEA). These experiments of ferromagnetic discs of NiMnSb have shown interesting behaviour [Loub 09]. Based on these results a non-volatile memory device can be envisioned [Pige 10a, Pige 10b].

4.1 Ferromagnetic resonance spectrometer

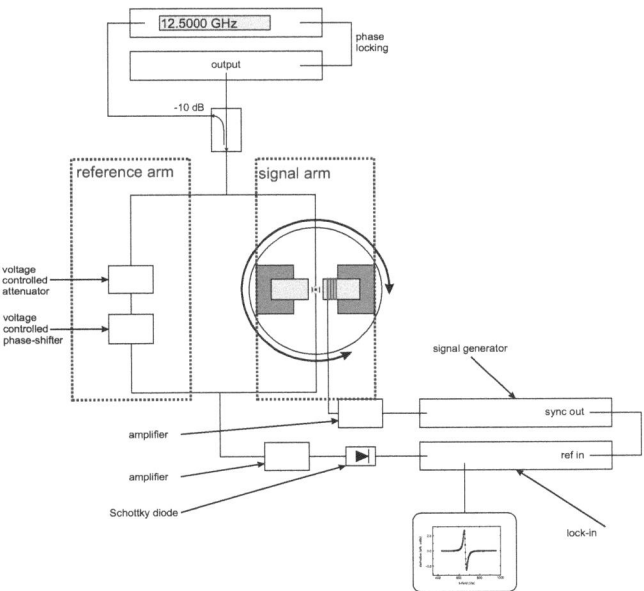

Fig. 4.1: Schematic of the used fmr spectrometer: the signal is divided using a splitter/combiner in a reference arm and a signal arm which are afterwards recombined by using a second splitter/combiner; the RF-amplifier is a broad-band amplifier with a gain of 43 dB; the modulation frequency is 156 Hz

Fig 4.1 shows a sketch of the used microwave spectrometer setup. Much in contrast to the widely spread use of a rectangular waveguide, an on-chip geometry is used. For the measurements presented in the following the coplanar waveguide (CPW) geometry is used. This geometry offers in contrast to other strip lines the advantage of designing impedance matched waveguides with different signal line widths on the same material. For CPW geometry the impedance can be tuned by adjusting the ratio of signal-line width to gapping width with respect to the dielectric constant of the used material [Poza 05].

Primary a broad-band GHz emitter is used. This emitter is phase-locked to an accuracy of 10 Hz using an external frequency counter. The line width of the emitted signal is in the order of a couple of hundred kHz and has a maximum output power of 15 dBm. The measurements of the magnetostatic modes are done using a small-band GHz emitter (10.000 GHz to 10.500 GHz) with a maximum output power of 0 dBm at a frequency of 10.368 GHz and a line width of the signal of only tens of kHz. This emitter is internally phase-locked.

A rotating magnet with a maximum field of 280 mT and a modulation coil wound to give the reference for the lock-in is used. The lock-in demodulates the voltage of a zero-bias negative Schottky-diode which is used to detect the high-frequency. In order to achieve a better signal to noise ratio an interference technique is used. By using a variable voltage-controlled attenuator and phase-shifter in the reference arm, a destructive interference can be adjusted and therefore the background signal can be reduced.

4.2 Ferromagnetic resonance measurements on NiMnSb

Ferromagnetic resonance is a key technique to determine magnetic properties of ferromagnets. Having the capability of full angle dependent measurements of the ferromagnetic resonance in-plane as well as out-of-plane for at least two frequencies offers the possibility to determine the in-plane and out-of-plane anisotropy constant, the saturation magnetization, the Gilbert damping factor G and the g-factor of the examined sample. The limitation in field strength of the used setup restricts to only in-plane FMR measurements. The in-plane measurements give access to in-plane anisotropy constants and the g-factor as well but the saturation magnetization itself can not be accessed, only the effective magnetization $4\pi M_{eff} = 4\pi M_s - K_u^\perp / M_s$.

4.2.1 NiMnSb grown on 001 InP substrate

As shown in 2005 by Koveshnikov et al [Kove 05] NiMnSb grown on (001)InP with an (In,Ga)As buffer exhibits a changing characteristic in the anisotropy. The FMR measurements indicated a thickness dependent changing from a uniaxial anisotropy with its easy axis along the [1$\bar{1}$0] crystal direction over a 'four-fold' back to a uniaxial anisotropy, with its easy axis now along the [110] crystal direction. The anisotropy field decreases from 400 Oe for the thinnest layer (8 nm) to 200 Oe for the 82 nm NiMnSb layer. Measurement of the line width of the absorption peak of these samples indicated a dependency to misfits which are along the [100] and parallel to the [110] direction which was in good agreement to performed TEM measurements which showed a edge-like defects aligned with the [100] and the [110] direction. The dimensionless damping parameter α was, for the thinnest layer measured (8 nm), in the low 10^{-3} [Hein 04] range which indicated the good quality even for layers with such an amount of defects.

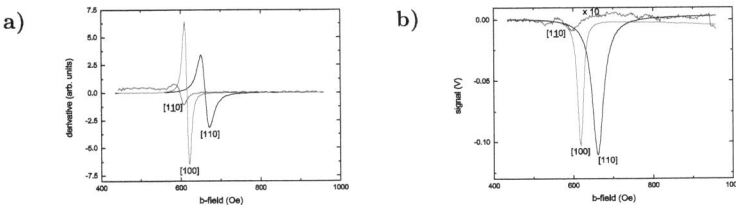

Fig. 4.2: A typical spectrum is shown (a) derivative as measured with lock-in technique; b) direct signal); the intensity fluctuations are due to the rotating of the external field and therefore having only for 0° and 180° a perpendicular orientation of the static field and the alternating GHz field; the graphs are recorded using different sensitivities of the lock-in

The presented measurements of the newly grown NiMnSb with changed flux compositions still show this thickness dependency of the anisotropy but decreased damping for thicker layers. NiMnSb sample without capping layers were measured and the results are presented. The figures 4.2 show typical spectra. Because of the use of a lock-in technique (the field is modulated) the derivative of the signal (fig. 4.2.a) is measured (fig. 4.2.b shows the direct signal). By rotating the magnet, the intensity decreases as shown due to a decreasing angle between the external static magnetic field and the hf-field. This is one major reasons for the absence of damping measurements around 90° because for this geometry the external field and the hf-field are aligned parallel and therefore only a small fraction of the alternating field is picked up. In fig. 4.3 angle dependent measurements

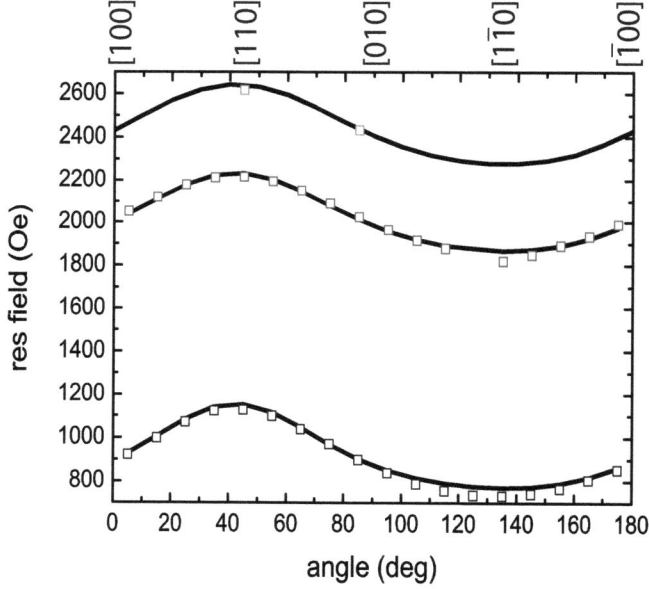

Fig. 4.3: resonance field versus in-plane angle for 10 nm NiMnSb for three different frequencies (black:8 GHz, red:12,5 GHz and green:14 GHz) and fits using parameters shown in tab. 4.1

for three different frequencies (from bottom to top:8 GHz, 12.5 GHz and 14 GHz) are shown. To fit the angle dependency and therefore find the parameters the following free energy function is used:

$$F_{100} = K_1(Cos^4(\varphi_4 - \varphi_M) + K_u Cos^2(\varphi_{uniaxial} - \varphi_M)) \tag{4.1}$$

with φ_M is the in-plane angle of the magnetization with respect to the (100) crystal axis and $\varphi_{4/uniaxial}$ the four-fold, the uniaxial respectively, angle with respect to 0°. Equation 4.1 is valid for a cubic symmetry with a superimposed uniaxial component.

4.2. Ferromagnetic resonance measurements on NiMnSb

For all angle dependent measurements done on NiMnSb layers in this section zero degrees correspond to the [100] crystal direction. In the plot shown above a uniaxial anisotropy is visible with an easy axis along the [1$\bar{1}$0] crystal direction and the hard axis along the [110] direction.

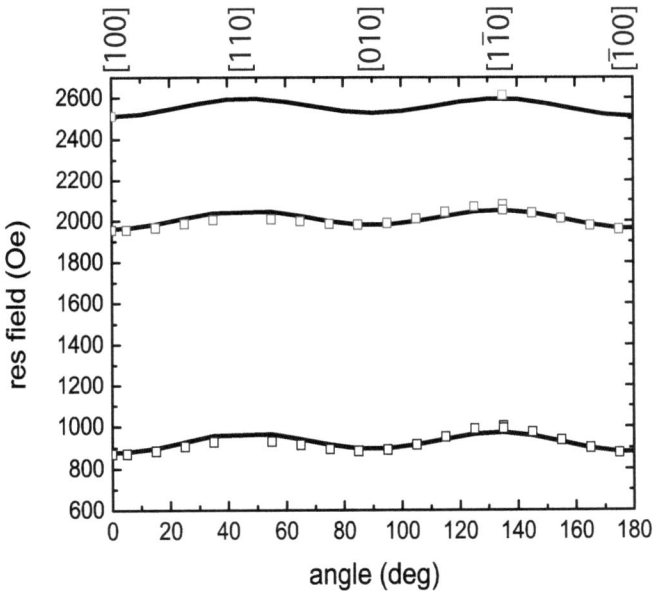

Fig. 4.4: Resonance field versus in-plane angle for 8 GHz (black), 12.5 GHz (red) and 14 GHz (green) (open boxes) and fitted angle dependencies (solid lines) using parameters from table 4.1

By increasing the layer thickness to 20 nm the uniaxial anisotropy vanishes and a 'four-fold' anisotropy becomes visible as shown in fig.4.4. Here the uniaxial anisotropy having a value of 17.8 Oe and the four-fold anisotropy is 20.5 Oe (fitting parameters are shown in tab.4.1). Compared to the 10 nm which was best fitted using -100 Oe for the uniaxial anisotropy and 12.7 Oe for the 'four-fold' anisotropy the uniaxial is decreasing and the 'four-fold' is increasing as reported earlier.

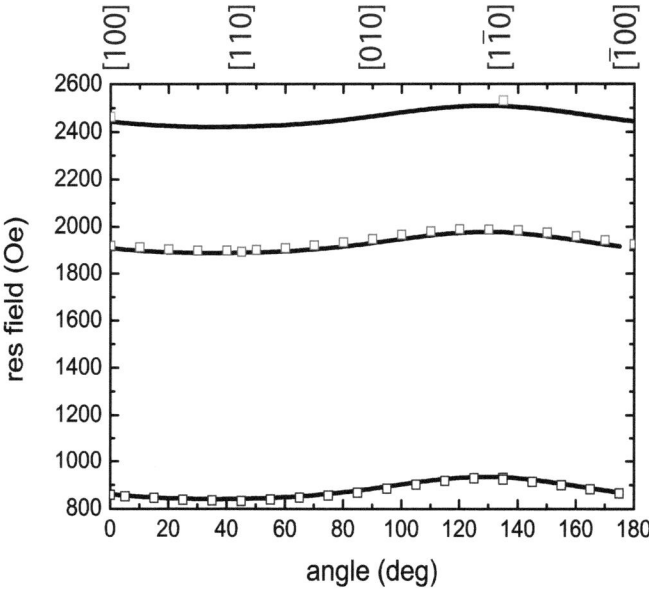

Fig. 4.5: The solid line is the analytically found fit (parameters according to tab. 4.1) and the open boxes are representing the measurements for three frequencies (black:8 GHz, red:12.5 GHz and green:14 GHz)

Fig.4.5 shows the angle dependent measurement of the in-plane resonance field of a 40 nm NiMnSb layer. The easy axis of the uniaxial anisotropy has changed from along the [1$\bar{1}$0] to parallel to the [110] crystal direction. TEM measurements indicated the presence of a preferential direction of misfits (along [100] and [110]). Indication of the presence of misfits can be seen in the line width measurements.

4.2. Ferromagnetic resonance measurements on NiMnSb

	$10 nm$	$20 nm$	$40 nm$
$2K_1/M_s (Oe)$	12.69	20.45	3.60
$2K_u/M_s (Oe)$	-100.77	17.76	24.88
$\Psi_4(°)$	-5.00	-0.97	-4.81
$\Psi_{uniaxial}(°)$	43.13	37.35	37.88
$M_{eff}(kG)$	7.57	7.70	7.89
g	2.04	2.04	2.04

Tab. 4.1: Parameters used for fitting the experimental data to the free energy function for NiMnSb layers with different thicknesses

In table 4.1 the fit parameters which are used to fit the angle dependent FMR measurements (solid lines in plots) using the free energy function 4.1 are shown.

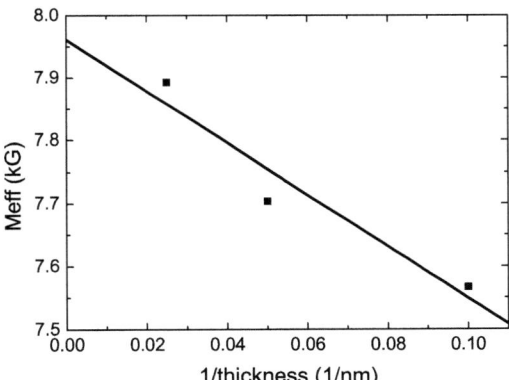

Fig. 4.6: Effective magnetization over inverse thickness is plotted. The decrease in the magnetization with decreasing thickness is consistent with the definition of the effective magnetization as the sum of the saturation magnetization and the out-of-plane uniaxial anisotropy which scales inverse to the thickness (see equ. 2.30)

The decreasing effective magnetization with decreasing thickness, as shown in fig. 4.6, can be understood by the definition of the effective magnetization (equation 2.30) as the sum of the saturation magnetization and the out-of-plane uniaxial anisotropy. Figure 4.7 shows a $1/d$ dependency for the in-plane uniaxial anisotropy. A negative uniaxial anisotropy field as for the 10 nm layer indicates a rotated easy axis by 90° (with respect to the [110] direction). This can easily be understood by recalling the equation for the free energy 4.1. Herein a negative factor K_u is equivalent to changing the symmetry of the trigonometric function from cosine to sine.

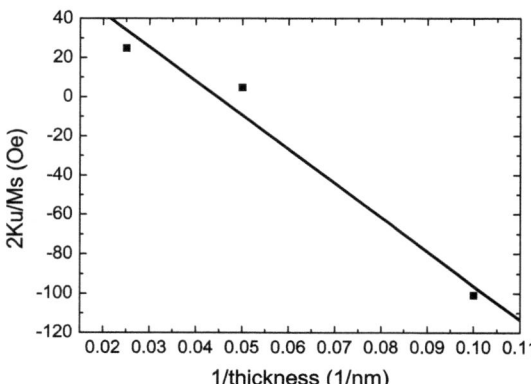

Fig. 4.7: The figure show the in-plane uniaxial anisotropy field strength versus the inverse thickness. A clear linear dependency (solid line corresponds to fit) indicates the expected 1/d dependency

By measuring the line width for more frequencies versus the in-plane angle φ an angle dependency of the damping can be calculated and in the same way an angle dependency of the inhomogeneous broadening can be plotted (fig. 4.8). In fig. 4.8.a) the damping of

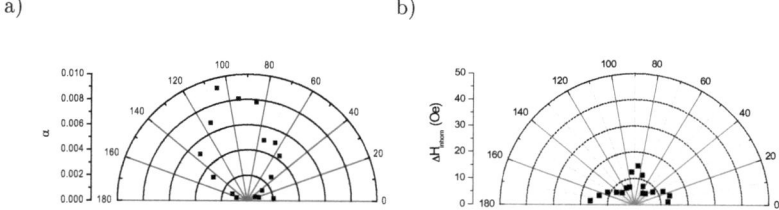

Fig. 4.8: In a) the damping of the 10 nm thick NiMnSb layer is shown and in b) the inhomogeneous broadening (the zero frequency offset) is shown

the 10 nm NiMnSb layer is shown in a polar plot. The texture in the damping shows a uniaxial character with high damping along the crystal axis (010) and low damping along the (100) axis. Fig. 4.8.b) shows the inhomogeneous broadening of the line width. A distinctive texture is visible in the plot which is an indication for two-magnon scattering [Hurb 98, Aria 99, Land 08]. The high and low values corresponds to the crystal axis, the cleaving edges respectively. In the angle dependent plot of the 20 nm layers is also a texture in the damping and offset visible (see fig. 4.9). The orientation of the high damping

4.2. Ferromagnetic resonance measurements on NiMnSb

a) b)

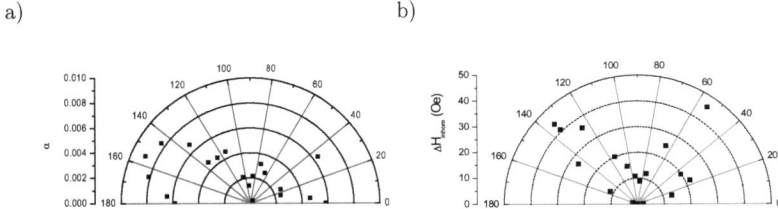

Fig. 4.9: On the left hand side (a) the damping is shown in a polar coordinate system whilst on the right hand side (b) the non-frequency offset (inhomogeneous broadening) of the 20 nm (001)NiMnSb layer is shown

changed from along the crystal axis (010) for the 10 nm layer to a high damping along the cleaving edge ($\bar{1}10$). A similar behaviour is visible in the inhomogeneous broadening. The orientation is rotated by 45° and the low offset is along the crystal directions (010) and (100).

a) b)

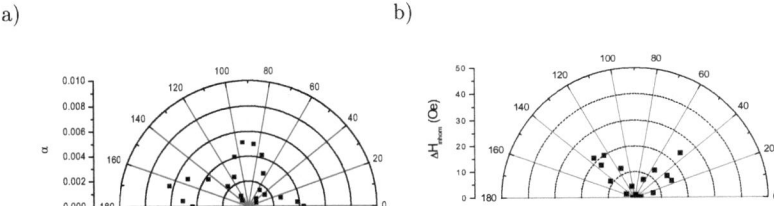

Fig. 4.10: Damping and inhomogeneous broadening extracted from the line width measurements of the 40 nm NiMnSb layer

The angle dependent damping parameter α of the 40 nm layer indicates a four-fold symmetry and do not show a two fold which is visible in the resonance field versus angle plots (fig. 4.5). For the 40 nm NiMnSb layer the texture in the offset is not rotated with respect to the offset of the 20 nm layer (see fig. 4.10b) but an additional high damping orientation becomes visible aligned with the crystal axis (010) (see fig. 4.10.a). This distinctive inhomogeneous broadening with this pronounced angle dependency is once again a hint for the presence of two magnon scattering processes. The inhomogeneous broadening plot 4.10 shows two minimums along the (100) and (010) crystal axis and two maximums along the (110) and ($1\bar{1}0$) cleaving edges. A clear δ ($\delta = \delta_{max} - \delta_{min}$) is observable. This is consistent with x-ray measurements of the sample which shows high crystal quality independent of the direction for the sample [Loch 10a]. The damping α of the 40 nm thick NiMnSb layer is shown in the following figure 4.11 for two different crystal directions.

The damping indicates a better crystal quality compared to the samples measured

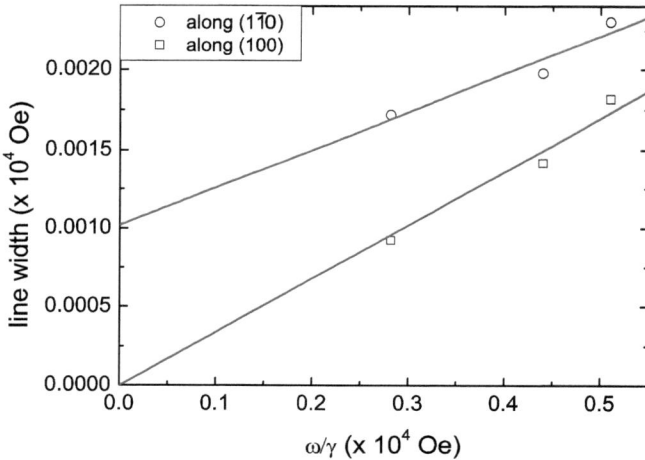

Fig. 4.11: Plot shows the line width over ω/γ for the 40 nm NiMnSb layer, the slope representing the damping, namely 4.74×10^{-3} along the (100) direction and 3.33×10^{-3} along the ($1\bar{1}0$) direction

earlier [Kove 05]. Early investigation of the NiMnSb grown on (001)(In,Ga)As buffer indicated low damping but the value of 10^{-3} was only found for thin (8 nm) NiMnSb layers. By increasing the thickness the line width of the absorption was increased as well and consequently the damping was enhanced. Nevertheless for the grown layers the low 10^{-3} value offers the possibility of fabricating devices for high frequency applications for which a good high frequency characteristic is essential. The good damping characteristics are one of the reasons why the measured magneto-static modes in NiMnSb sub-micrometer sized structures became observable (see chapter 4.4) Later in this work measurements of spin torque oscillators fabricated with NiMnSb as one ferromagnetic layer are presented.

4.2.2 NiMnSb grown on 111 InP substrate

In 2001 Wijs and co-workers initiated the use of NiMnSb layers grown along the [111] direction to avoid loss of spin polarization due to symmetry braking at interfaces and therefore offering the predicted 100% spin polarization [Wijs 01]. Therefore NiMnSb samples were grown on (111) InP substrate. The following section describes the FMR measurements of these layers. These substrates have a miscut of 2° along the $(11\overline{2})$ direction which is necessary to ensure a good growth start. Fig.4.12.a) shows an atomic force microscope (AFM)

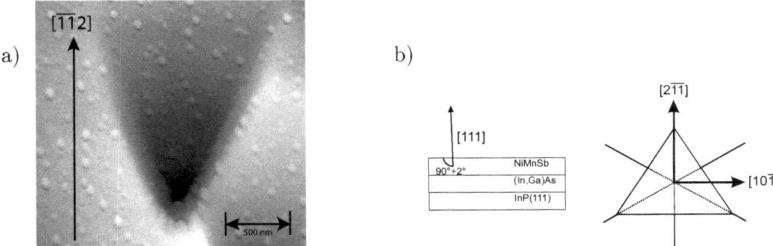

Fig. 4.12: a) Picture of an atomic force microscope (AFM) scan of the surface of a 10 nm thick NiMnSb layer grown on (111)(In,Ga)As buffer on (111)b InP substrates; b) drawing of the (111)NiMnSb sample with its triangular shape

scan of the surface of the (111)NiMnSb layer. In the scan a triangular dislocation is visible which is aligned with the miscut of the substrate, therefore mirroring the steps of the substrate. The dislocation is over 150 nm deep and therefore starting in the buffer which is 300 nm thick. More likely the defect is introduced by the substrate and vanishes by increased layer thickness. Similar to the NiMnSb layers grown on (001)(In,Ga)As buffer the half-Heusler alloy grown on (111)(In,Ga)As shows a uniaxial anisotropy. Fig.4.13 shows an angle dependent measurement of the resonance field for a 10 nm NiMnSb grown on (111)InP substrate. This uniaxial anisotropy is somehow surprising, because the free energy is predicting no anisotropy for this measurement geometry. Equation 4.2 is the free energy for a crystalline ferromagnet with cubic symmetry grown in (111) direction.

$$F_{111} = \frac{1}{3}K_4 - \frac{2}{3}K_4 sin^2\theta + \frac{7}{12}K_4 sin^4\theta + \frac{1}{3}\sqrt{2}K_4 sin(3\varphi) \cdot cos(\theta) \cdot sin^3(\theta) \qquad (4.2)$$

Taking $\theta = 90°$, which corresponds to measure in the (111)-plane, the upper equation results in an angle independent constant value. Much in contrast fig. 4.14 shows a uniaxial anisotropy with its easy axis aligned with the $(\overline{11}2)$ crystal direction. This is identically to the direction of the miscut. Therefore one can assume the symmetry breaking due to the miscut is causing a uniaxial anisotropy overlapping the homogeneous anisotropy according

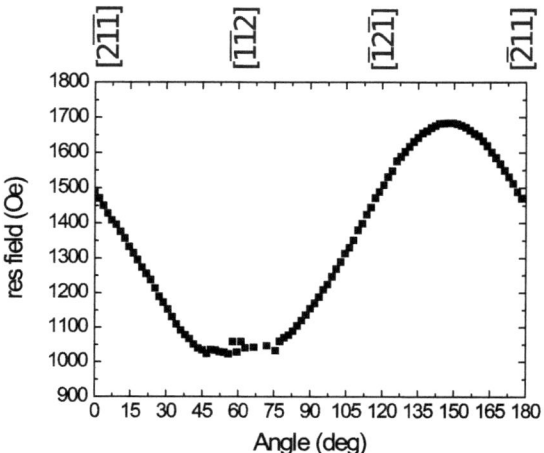

Fig. 4.13: Resonance field versus angle for a 10 nm thick NiMnSb layer grown on (111)(In,Ga)As buffer, a pronounced uniaxial anisotropy is visible with its easy axis along the ($\overline{1}\overline{1}2$) crystal direction

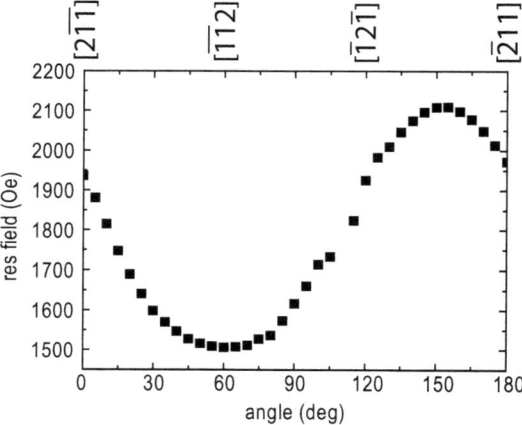

Fig. 4.14: An angle dependency (resonance field versus in-plane angle with respect to the ($2\overline{1}\overline{1}$) crystal direction) of 20 nm NiMnSb layer grown on (111) InP is shown

to the formula. Consequently a new empirical free energy function can be assumed:

$$F_{111_{miscutted}} = F_{111} + K_2 cos\varphi \tag{4.3}$$

4.2. Ferromagnetic resonance measurements on NiMnSb

To verify the miscut is the origin of the uniaxial anisotropy a 20 nm thick (111)NiMnSb layer is measured as can be seen in fig. 4.14. The uniaxial anisotropy is visible as well and its easy axis is align along the ($\bar{1}\bar{1}2$) direction comparable to the 10 nm layer. A decreasing anisotropy strength by increasing layer thickness (see tab.4.2) is indicated and verified by measuring a 40 nm layer of (111)NiMnSb as shown in fig.4.15.

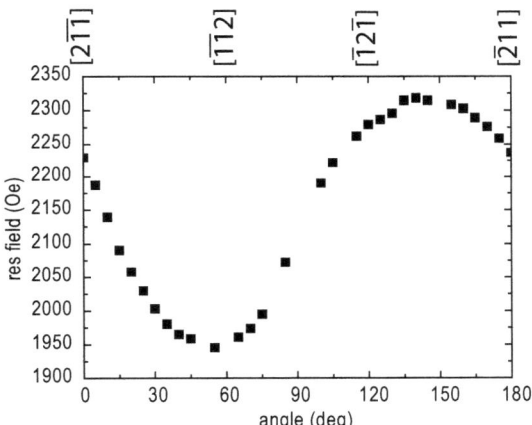

Fig. 4.15: Angle dependent measurement of the resonance field (40nm (111)NiMnSb) with a pronounced uniaxial anisotropy similar to all previously measured NiMnSb layers grown on (111)(In,Ga)As

The direction of the anisotropy remains aligned with the miscut, whilst the strength is decreasing with thickness. Figure 4.16a) and b) show the anisotropy fields plotted over

	40 nm	20 nm	10 nm
$2K_u/M_s$ (Oe)	335	560	700

Tab. 4.2: table of anisotropy strength for the (111)NiMnSb grown on (111)InP substrates with (In,Ga)As buffer for different layer thicknesses

the thickness and the inverse thickness (bottom, top x-axis respectively).

The best fit straight line for both cases is fitting appropriate. One reason is the difference in the samples. The measured NiMnSb layers are partly capped (20 nm layer is capped by Ti/Au) and partly uncapped. This results in the presence or absence of an oxide layer at the surface. An oxide layer is reducing the magnetization and therefore the thickness of the layer. A second reason is the difference in the growth conditions. While the 40 nm layer is grown as almost the first layer of NiMnSb on (111)(In,Ga)As the 20 nm layer is grown after improvement in the growth condition and therefore offering better

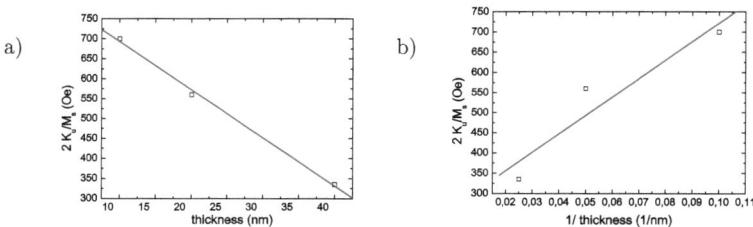

Fig. 4.16: Anisotropy field strength over layer thickness (a) and over inverse layer thickness (b) is shown. The red line is a linear fit, which fits adequately well to both datasets

crystalline quality, less dislocations. Figures 4.17 a) and b) show the line width of the absorption peaks versus the angle of the static magnetic field in a polar plot for the 20 nm and the 40 nm layers.

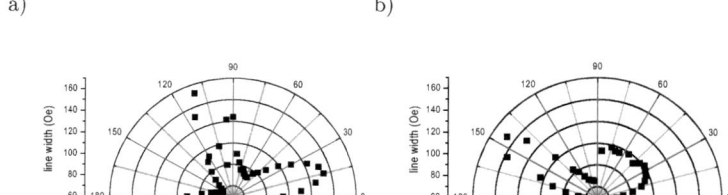

Fig. 4.17: Line width of the absorption peaks for the angle dependent in-plane FMR measurement of the NiMnSb layers which are grown on (111)(In,Ga)As, a) 20 nm thickness and b) 40 nm thickness

In the angle dependent line width plot for the 40 nm thick layer a fourfold behavior is visible with the biggest value along 150°. The line width angle dependency of the 40 nm NiMnSb layer shows a six-fold symmetry with 3 maximums (along 30°, along 90° and along 150° (see fig. 4.17.b)). According to the free energy equation 4.2 it can be seen for an out-of-plane angle $\theta \neq 90°$. This is realized by measuring with a slight tilt which is not effecting the anisotropy field but effecting the line width angle dependency.

Summary The ferromagnetic resonance measurements of the (001)NiMnSb samples indicated the high quality of the grown layers. A small in-plane anisotropy is visible and showed the well known behaviour from uniaxial for thin layers over four-fold to a 90° rotated uniaxial anisotropy for thick layers [Kove 05]. Whilst for damping measurements of the reported layers a low damping could only be measured for thin layers (8 nm), the measured layers in this thesis showed remarkable low damping for 40 nm thick layers.

4.2. Ferromagnetic resonance measurements on NiMnSb

As a second part NiMnSb grown on (111)(In,Ga)As were measured. A more pronounced thickness dependent uniaxial anisotropy was visible for all samples with a direct dependency to the direction of the miscut in the used substrate. The damping is higher compared to (001)NiMnSb samples.

4.3 Spin pumping

Tserkovnyak and co-workers [Tser 02] theoretically described the experimental observation of Mizukami et al. [Mizu 01a, Mizu 01b] for capped Permalloy layers by the use of spin-pumping and spin-sinking concept. In this section FMR measurements of NiMnSb layers grown on (001)(In,Ga)As with different capping layers are presented. The measured NiMnSb layers have thicknesses of 5, 8 and 10 nm. All FMR measurements are done in-plane with the external field applied along the (100) crystal direction. This direction offers the smallest line width for NiMnSb layers (see chapter 4.2.1). Fig.4.18 show the damping of the NiMnSb capped with different non-magnetic metals.

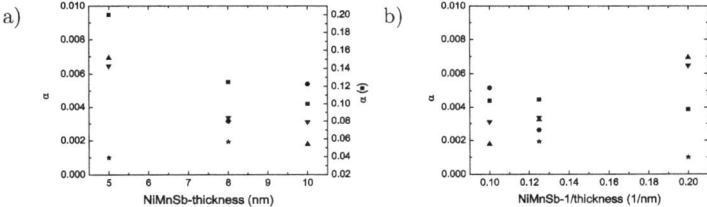

Fig. 4.18: Damping parameter α for different capping materials on NiMnSb with different thicknesses (✶:Cu, ▲:Pt, ▼:Ta, ●:MgO, ■:Ru);a) damping over thickness;b) damping over the inverse thickness for all capping materials

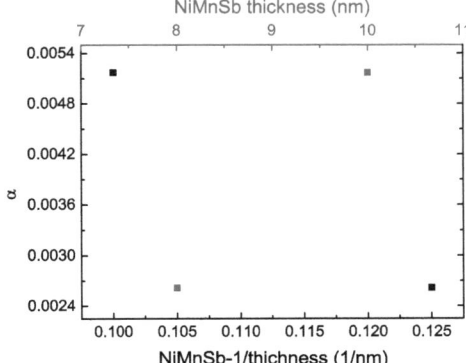

Fig. 4.19: Pumping using MgO as capping layer; the increase with increasing thickness is contradicting the spin-pumping concept

Figure 4.19 shows the damping parameter α over the inverse thickness of the NiMnSb layer capped with MgO. The increasing damping for thicker materials is an artefact with its origin not in the spin pumping mechanism. During the measurements the intensity of the signal was weaker and noisier compared to the other examined materials. The lower intensity can be seen as is a hint of less ferromagnetic material because the intensity is direct proportional to the volume of the ferromagnet. An explanation can be the presence of a dead layer when depositing MgO on the NiMnSb layer [Zhou 06]. Similar behaviour has been reported for NiMnSb covered with Mo and Au by Hordequin and co-workers [Hord 98]. A possible explanation might be that the Oxide in the MgO influences the NiMnSb which is in direct contact with the MgO and the spin pumping is prohibited. This can be an explanation for the absence of a 1/thickness dependency of the damping. For the damping of the NiMnSb layer using Cu as capping material (see fig. 4.20) an almost thickness independent damping behaviour is visible (the damping decreases from 19×10^{-3} to 10×10^{-3}). This is consistent with Copper being a weak spin sink and therefore

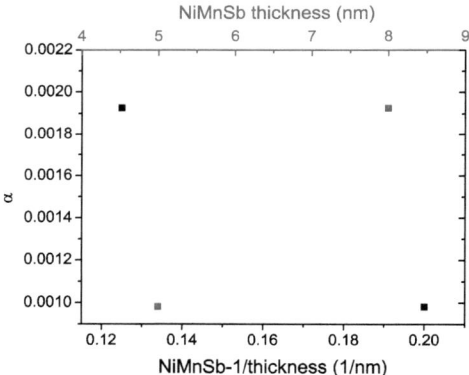

Fig. 4.20: NiMnSb with Cu as capping layer comparable to MgO

influences the damping only little [Mizu 01a, Mizu 01b]. On the other hand the Cu layer has a high conductance which influences the relaxation time [Mede 78] which is proportional to Z^4 [Abri 62].

The slope of the linear fits differs for the materials between 0.035 for Ta and 0.051 for Pt. Damping measurements of NiMnSb covered with Ru showed a negative dependency of the damping to the inverse thickness. During the last time a lot of measurements of the magnetoresistance of different sputtered spin-valves were reported in which an additional Ru layer increases the magnetoresistance. The measurements show a spin sinking behaviour comparable to the copper layer which has a large spin diffusion length and therefore a negligible influence to the damping due to spin pumping into the metal.

4.3. Spin pumping

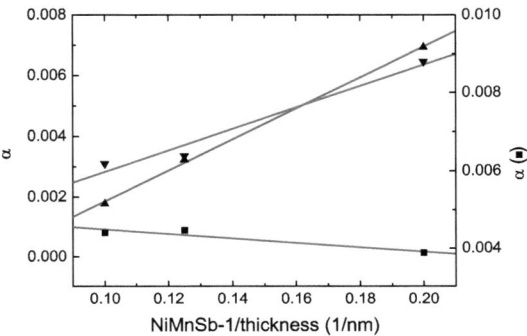

Fig. 4.21: Damping over the inverse thickness of the NiMnSb layer for Ta, Pt and Ru (▲:Pt ,▼:Ta, ■:Ru) as capping metals

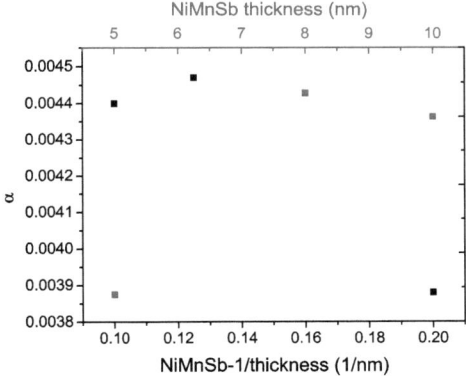

Fig. 4.22: the figure above shows the damping parameter over the inverse-thickness (lower x-axis; over the thickness, upper x-axis) for NiMnSb layer capped with Ru

Using tantalum or platinum gives a behaviour of the extra damping due to spin pumping as expected from theory.

The slope of 0.0534 for the best fit straight line for Py covered with Ta (see fig. 4.23.a) is bigger than the 0.035 for the NiMnSb covered with the same material (see fig. 4.23.b). In figure 4.24 both damping over inverse thickness for Py and NiMnSb is shown. The difference in slope as well as the difference in damping is visible.

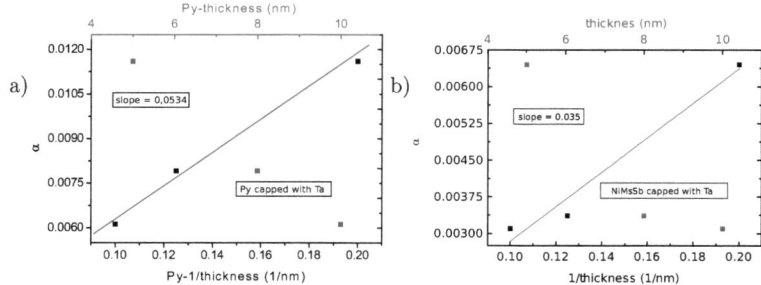

Fig. 4.23: a) shows a the damping of a Permalloy (Py) layer covered with Ta for different thicknesses, the linear fit results in a slope of 0.0534 while in b) NiMnSb layer capped with Ta is shown and the corresponding best linear fit having a slope of 0.035

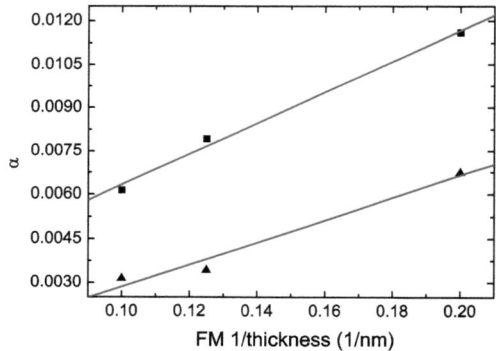

Fig. 4.24: The figure above shows a comparison of the damping parameters for Py and NiMnSb layers covered with Ta (▲:NiMnSb, ■:Py)

In fig. 4.25 the damping over the inverse thickness of the NiMnSb capped with 30 nm Platinum is shown. The increased slope compared to the NiMnSb capped with Ta is consistent with measurements done using Py as ferromagnet [Mizu 01a]. Mizukamie et al observed a dramatic difference between the thickness dependent damping of the permalloy capped with Ta and Pt. This was attributed to the different atomic number and $1/\tau_{sf}$ scales with z^4 [Abri 62].

4.3. Spin pumping

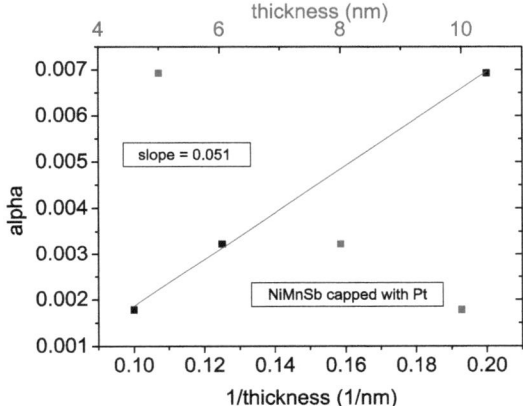

Fig. 4.25: Damping of the NiMnSb layer capped with 30 nm Pt over inverse thickness

Summary

The measurements of spin-pumping from the ferromagnetic NiMnSb being in contact with different non-magnetic layers is an essential step in order to realize spin-torque devices. Thickness dependent measurements of the damping of NiMnSb in contact with Cu showed no significant dependency and therefore indicates the NiMnSb-Cu interface being of good quality. The measurements of various other gapping materials showed similar thickness dependent behaviour as previously reported experiments using Permalloy as ferromagnet.

4.4 Magnetostatic modes

To verify that the high frequency characteristics of the NiMnSb remain good after processing, sub-micrometer sized elements were fabricated. In order to improve the measurement sensitivity, the structures were fabricated lithographically with a coplanar waveguide patterned on top with an insulating interlayer. This allows to exactly position the structures under the coplanar waveguide and therefore ensures a maximum FMR intensity. Different pillar shapes, namely squares, discs and stripes, were measured. The stripes were aligned with the long side parallel to the signal line of the coplanar waveguide and the field was swept in the same direction.

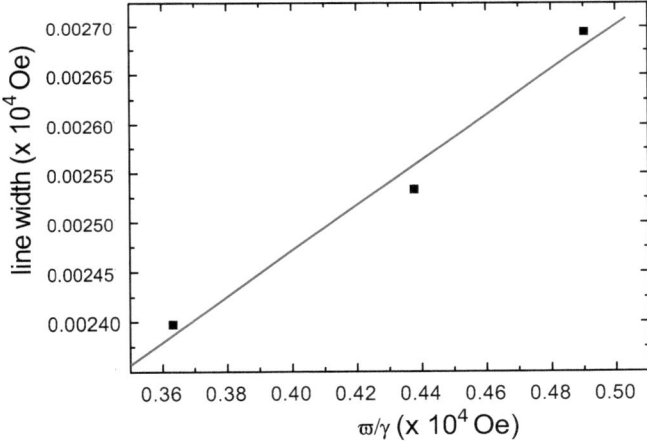

Fig. 4.26: Line width of absorption measured for rectangular NiMnSb elements (600 times 1200 nm², 40 nm thick) with the external field applied parallel to the long side of the stripes, the solid red line indicates the best linear fit and therefore results in a damping α of 2.3×10^{-3}

Figure 4.26 shows the line width of the absorption peak for 10.368, 12.5 and 14 GHz. From the slope of the linear fit (solid red line), the damping parameter α is 2.3×10^{-3}. This value is comparable to the damping measured for unprocessed layers (see chapter 4.2.1).

In the spectra shown in fig.4.27.a) and b) additional absorption peaks become visible. This has been reported in studies of sub-micrometer sized permalloy structures using Brillouin light scattering (BLS) [Jorz 99, Rous 01] and is attributed to standing spin waves which are established when confining the lateral dimensions. The most intense peak corresponds to the uniform 1/1 mode and the higher modes correspond to modes with different wave numbers k_x and k_y. In the figure 4.27.a) the ferromagnetic resonance for two squares with dimensions of 10 times 10 μm^2 located symmetrically besides the

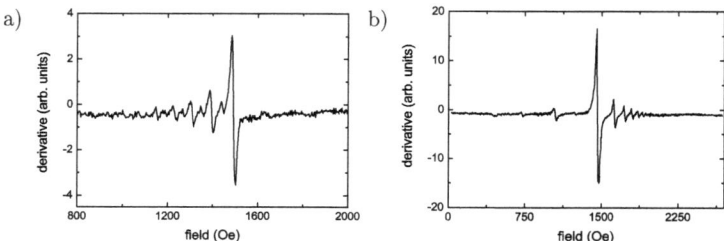

Fig. 4.27: The plots show the spectra of a) two 10 time 10 μm^2 squares and b) an array of 10 times 10 discs with a diameter 2 μm

signal line is shown. Because of the symmetry and the resulting absence of an additional anisotropy field from the shape the measurements were used as reference measurements. In fig. 4.27.b) the spectra of an array of 10 times 10 discs with a diameter of 2 μm is shown. Both graphs show higher resonance whilst the position of the main peak is nearly not influenced. The reason is in both cases the lack of shape anisotropy which would influence the internal field and hence the resonance condition (see equ.2.24 and equ.2.32). In fig.4.28 the spectra of stripes with different aspect ratios are shown.

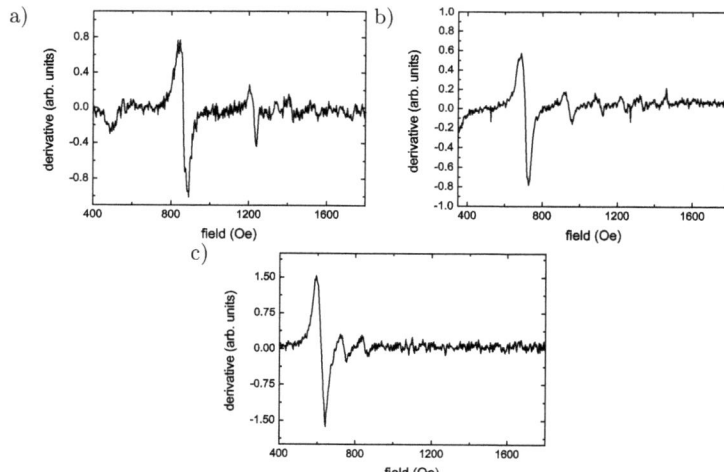

Fig. 4.28: Magnetostatic modes for arrays of stripes with different aspect ratio: a) 600 x 1200 nm^2, b) 600 x 2000 nm^2 and c) 600 x 3400 nm^2 stripes

Here the main peak is shifted systematically to lower fields for stripes with higher aspect ratios created by changing the length. A comparison of the different spectra is shown

4.4. Magnetostatic modes

in fig.4.29. The shifted main peak can, analogue to the considerations for the squares and

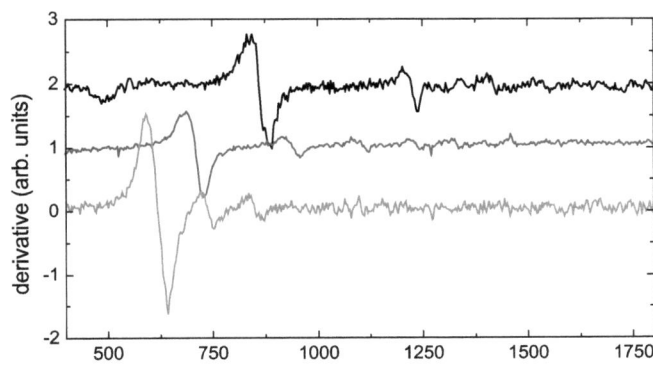

Fig. 4.29: Comparison of spectra for stripes with different aspect ratio (curves are offset by 1 for clarity)

discs, be explained with the shape of the elements and the additional anisotropy. In table 4.3 the anisotropy fields for stripes of NiMnSb (with a thickness of 40 nm) with different aspect ratios are listed.

Besides the shifted main peak higher modes become visible as well. An obvious depen-

thickness	lateral dimension	anisotropy field
40 nm	600 x 1200 nm^2	288 Oe
40 nm	600 x 2000 nm^2	417 Oe
40 nm	600 x 3340 nm^2	498 Oe

Tab. 4.3: Calculated anisotropy fields for rectangular particles of NiMnSb using 0.7 T as saturation magnetization

dence of the spacing of the peak position and the aspect ratio is visible, namely the lower the ratio the bigger the spacing. Walker and co-workers established a model to calculate the magnetostatic modes for spheres [Walk 57]. Later Bryant and co-workers [Brya 93] extended the method for rectangular elements.

Here, in order to verify the presence of magnetostatic modes, simulations with OOMMF ([Dona 99]) have been done. OOMMF can simulate the time dependent response of a magnetic system. The dimensions of the ferromagnetic elements are fed in the simulation program in combination with material parameters (damping, saturation magnetization, anisotropy). A static magnetic field which corresponds to the resonance field of the main mode is applied in the same direction as the external field in the measurements and the static magnetization of the element is tilted out of its equilibrium state by applying a Gaussian pulse (see fig.4.30.a).

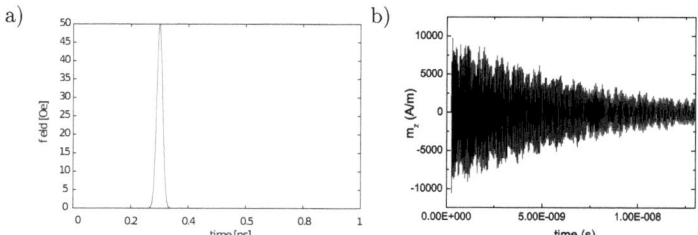

Fig. 4.30: Above an illustration of the used pulse (a) which is used in the simulation to start the relaxation process in the elements and (b) the resulting, in time decaying, magnetization of the element is shown

The pulse is centred at the time 3×10^{-13} seconds and has an amplitude of 50 Oe. By applying the pulse all modes are excited simultaneously and start to relax in their equilibrium state. Because of the difference in frequency for the different modes the time domain magnetization (as shown in fig. 4.30.b) exhibits a superposition of all frequencies. The decay in the figure is due to damping which was simulated using an damping factor α of 2.3×10^{-3}. By performing a fast Fourier transformation (FFT) the frequencies can be determined. Afterwards the time domain data has to be recalculated to be field domain. Because of using the main peak to adjust the applied field to be equal to the field corresponding to the main absorption in the measurement a recalculation using the Kittel formula [Kitt 49] is done.

4.4. Magnetostatic modes

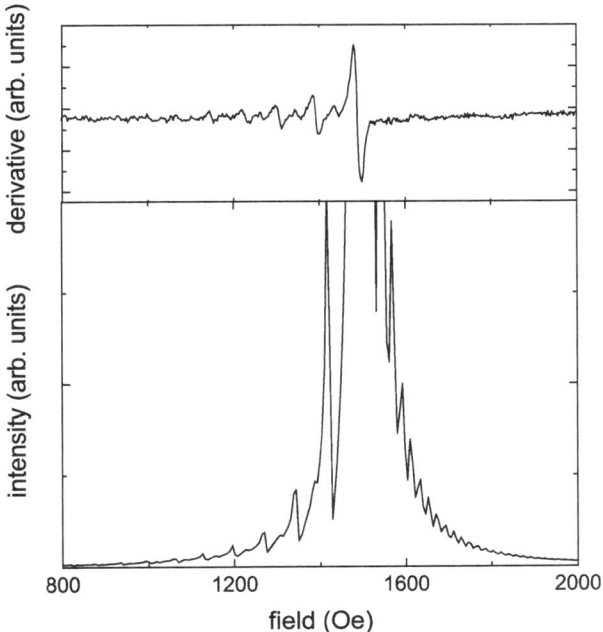

Fig. 4.31: Measurement and FFT of the simulated data for the 10×10 μm^2 squares

Fig. 4.31 shows a measured FMR spectrum (upper graph) and the calculated Fourier transformation of the simulation performed with OOMMF (lower graph). Based on the main peak the resonances on the low field side of the main peak correspond well with the measurement. The resonance on the high field side which are visible in the simulation and are not present in the measurement can have their origin in a small variation of the shape due to processing issues. By increasing the size of the element the peaks narrow more and more until the limit case of a infinite plane. In 2006 Bailleul and co-workers [Bail 06] simulated magnetostatic modes for permalloy squares with 3 μm sides which indicates the broadening of the resonances for smaller elements.

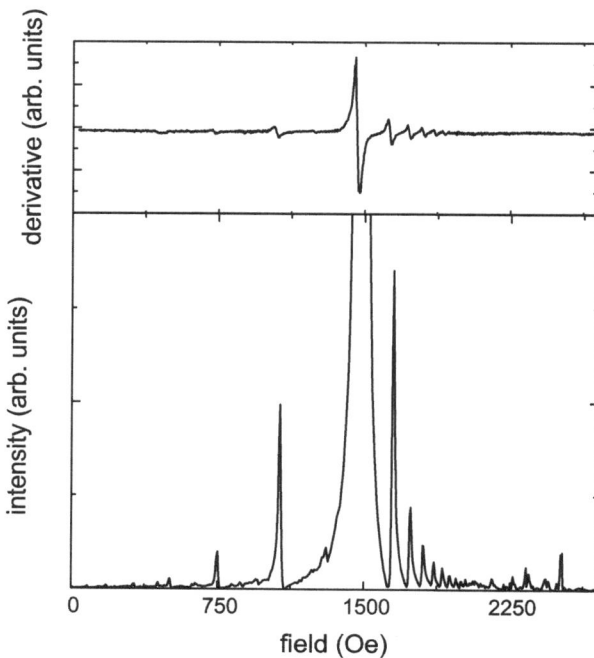

Fig. 4.32: For the discs (measurement upper graph and FFT of simulation lower graph) the resonances correspond well

In fig. 4.32 the calculated FFT of the simulation of circular discs with a diameter of of 2 µm analogue to the measured specimen in combination with the measurement is shown. The simulated peak position corresponds well with the measured absorption peaks of the magnetostatic modes. The used parameters are 9.07 kG saturation magnetization and a uniaxial anisotropy field of 90 Oe with its easy axis perpendicular to the applied field. As a result of a angle dependent in-plane FMR measurement of the unprocessed layer this uniaxial anisotropy has to be considered in the simulations as well. In the measurement as well as in the simulation are resonances with significant less intensity visible between the main peak (1490 Oe) and the first low field resonance (1100 Oe), namely at 1290 Oe.

4.4. Magnetostatic modes

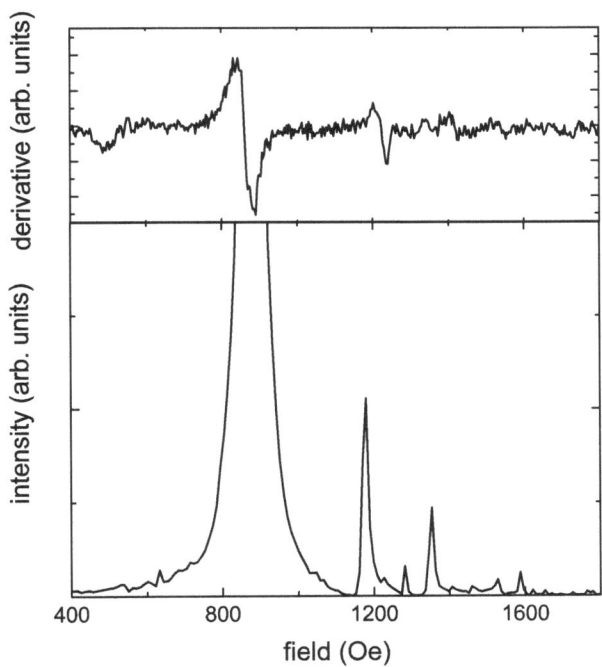

Fig. 4.33: The upper graph shows the measured spectrum of an array of 10 times 10 600×1200 nm² stripes while the lower graph shows the corresponding FFT of the simulation

Fig. 4.33 to 4.35 show the measured spectra and the corresponding FFT of the simulation for stripes with different aspect ratios. All shown simulations correspond well with the measured spectra. Fig. 4.33 shows the result for the 600×1200 nm² stripes. The main peak is shifted to lower field due to the upcoming shape anisotropy.

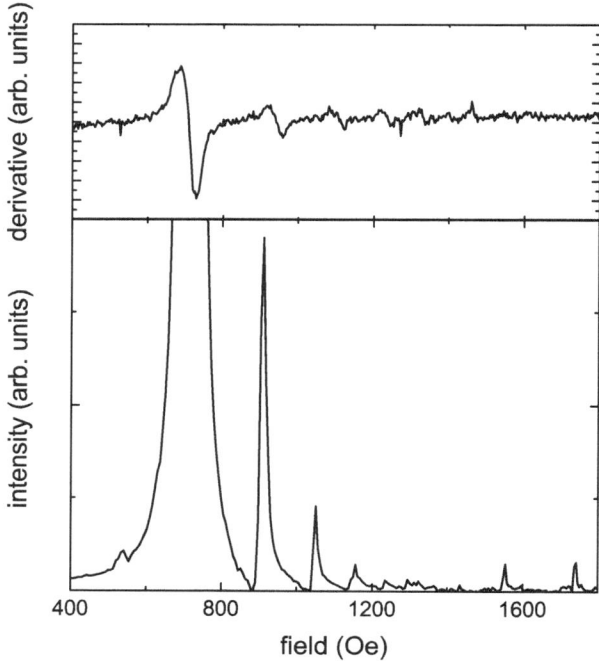

Fig. 4.34: The lower graph show the FFT of the OOMMF simulation of an array of 10 times 10 stripes with dimensions of 600×2000 nm² and is in very good agreement to the measurement of the FMR as can be seen in the upper plot

Fig. 4.34 shows the measured FMR and the FFT of the simulated spectrum. The main peak is shifted further to lower fields due to a higher shape anisotropy based on an increased aspect ratio. In the FFT the higher mode peaks correspond well with the absorption peaks on the high field side of the main peak in the measurement.

4.4. Magnetostatic modes

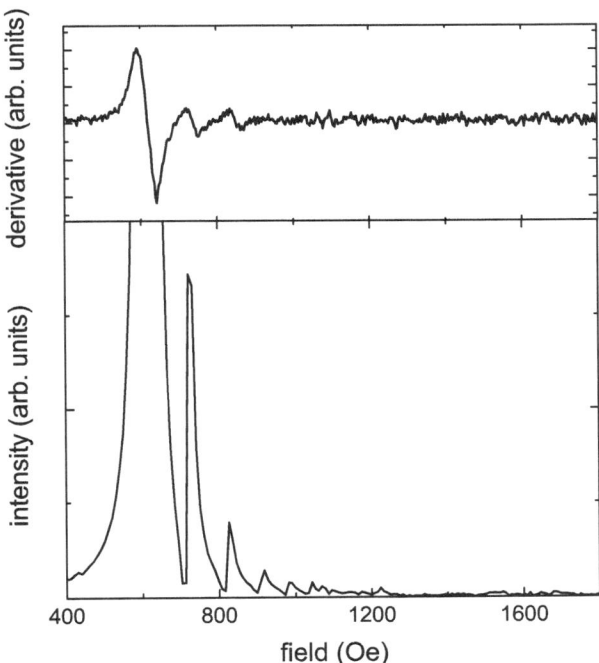

Fig. 4.35: An array of 10 times 10 stripes (600×3400 nm^2) is measured using FMR (upper graph) and compared to simulations done with OOMMF (lower plot)

An other indication for the presence of magnetostatic modes is the good agreement of the simulation and measurement for the 600×3400 nm^2 stripes as shown in fig.4.35. The lowest field position for the main peak compared to the spectra for stripes with lower aspect ratio is explainable with the shape anisotropy as mentioned above.

Summary

The measurement of magnetostatic modes in fabricated sub-micrometer sized elements of NiMnSb allowed an indirect verification of the remaining good quality. The low damping as could be measured directly for the 600×1200 nm^2 emphasize the use of NiMnSb in nanometer sized devices for spin-torque measurements.

4.5 Summary of hf measurements of NiMnSb

Sections 4.2 to 4.4 indicate NiMnSb to be a good candidate for high frequency devices. NiMnSb offers very low damping in combination of an in-plane four-fold anisotropy with a superimposed uniaxial anisotropy which changes its easy axis by 90° for layers with thicknesses between 10 and 40 nm (chapter 4.2). The low damping is a major necessity for high-frequency devices such as spin-torque oscillators. In chapter 4.3 spin pumping experiments using NiMnSb show a comparable pumping as has been reported for Py. The measurements indicating the sensitivity of the pumping to the sputtered process, the interface quality respectively. At last the chapter 4.4 proofs the remaining good quality of NiMnSb after processing small structures. The damping is in the same order as for unprocessed layers while the confined lateral dimensions cause the presence of magnetostatic modes. OOMMF simulations are presented to verify the higher visible absorption peaks have their origin in magnetostatic modes.

4.6 Spin-torque devices using NiMnSb

The measurements presented so far indicate the high crystalline quality and therefore a good high frequency behavior of NiMnSb. The following chapter presents the transport measurements performed on NiMnSb based pseudo-spin-valves. First magnetoresistance measurements of GMR devices using NiMnSb will be presented and afterwards the focus is on spin-torque measurements. The current densities necessary to switch the magnetic configuration of the spin-valve is as small as the densities of state-of-the-art devices and the high frequency emitted by the device, spin-torque oscillators respectively, have a q-factor which is orders of magnitude higher than the highest reported q-factor while competing in other parameters with the state-of-the-art GMR devices [Li 09, Mang 06, Rave 06] and almost comparable to TMR based spin-valve structures [Huai 04].

4.6.1 magnetoresistance

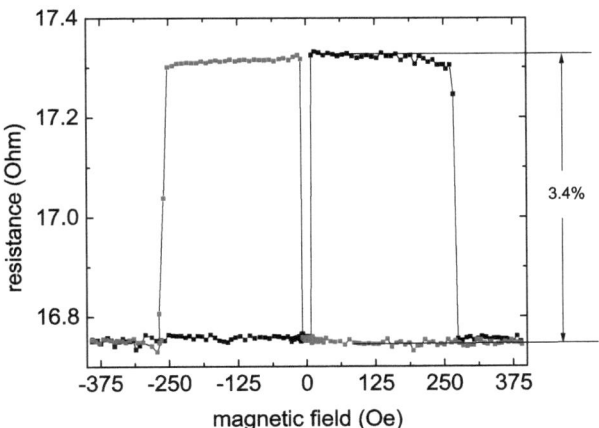

Fig. 4.36: Room temperature magnetoresistance measurement of a NiMnSb based pseudo-spin-valve ((40 nm)NiMnSb/(10)Cu/(6)Py) measured in an external field swept along the long axis of the elliptically shaped pillar (80 x 200 nm^2)

In fig. 4.36 a magnetoresistance measurement of the previously described pseudo-spin-valve is shown. A parallel configuration was established by applying a strong negative field parallel to the long axis of the elliptical pillars. The low resistance value indicates a parallel state. By increasing the field from high negative values to zero the slope of the measured resistance is almost zero. This is a major indication for being aligned with the easy axis defined via the shape of the pillar. A misalignment of the field and the anisotropy would manifest by a rotation of the magnetization from along the external field to along the easy magnetization direction with decreasing external field and therefore in a slope on the resistance curve. The switching visible after crossing zero external magnetic field is proof of a coupling between the two ferromagnetic layers which is significant lower than the coercive field of the softer ferromagnet. An increase to the maximum resistance value of 17.3 Ω without steps in the switching representing the device being in a single-domain state. After reaching a threshold field of 280 Oe the pseudo-spin-valve established the parallel configuration again. The calculated GMR ratio of 3.4 % is a proof of the not 100 % spin polarization of the NiMnSb polarization layer. The red curve shows the back sweep with the same behaviour in the manner of after-zero switching and zero slope.

Fig. 4.37 shows a measurement done on the same device at low temperature. Due to the temperature dependence of the magnetization direction the GMR ratio increases from 3.4 % to 7.5 %. In [Caba 98] a GMR ratio of 7.2 % is reported for a comparable pseudo-spin-valve. In the paper a 'normalized' ratio is reported which is falsified by contact resistances. Taking into account these resistances the GMR ratio would decrease further more. Much in contrast to the presented measurements by Caballero and co-workers the magnetoresistance measurements in this thesis are not 'corrected' but are presented as measured (no contact resistance has been extracted). Both measurements, the measurements reported and the measurements shown here, indicating a major difficulty using (001) NiMnSb for spin-transport devices, namely the loss of spin polarization at the interfaces.

The measurements of the GMR ratio for room temperature and for low temperature indicating the potential of the half-Heusler alloy. The GMR ratio is still low but by the use of (111) NiMnSb the interface effects might be negligible or significant lower.

4.6.2 current induced switching

By biasing a current above a certain threshold it is possible to change the configuration of a spin-valve from the parallel to the antiparallel state and vice versa.

Figure 4.38 shows a current induced switching measurement at room temperature. The external field was off during these measurements. One can distinguish two resistance state which correspond to the values measured by varying the external field. The parabolic behaviour of the resistance is caused by Jule heating effects. By applying a high negative current an antiparallel state (AP) was established at the beginning of the measurement. Afterwards the current was decreased constantly. At a current of +2.68 mA the applied spin-torque was high enough to switch the configuration from antiparallel to parallel. A small resistance value indicating this parallel state (P). By increasing the current further the configuration of the pseudo-spin-valve is not changing. The shoulder on the negative current side around zero might be an indication of an additional configuration state of the

4.6. Spin-torque devices using NiMnSb

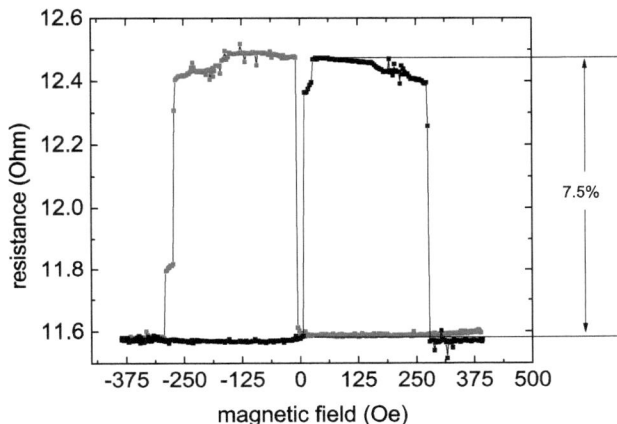

Fig. 4.37: Magnetoresistance measurement for the same device done at low temperature (80 × 200 nm^2; (40)NiMnSb/(10)Cu/(6)Py)

two ferromagnetic layers. By reducing the current the spin-torque which acts to stabilize the parallel state is decreased and another configuration can appear. By decreasing the current further to negative values the magnetic configuration changes again from parallel to antiparallel at a threshold current of -2.81 mA. The corresponding current densities can be calculated using the equation:

$$j_{C\pm} = A_{ellipse}/I_{threshold} \; [\frac{A}{cm^2}] \tag{4.4}$$

The resulting values for this 80 times 200 nm^2 pillar are $2.13\times10^7 \frac{A}{cm^2}$ for switching from antiparallel in the parallel state and $2.24\times10^7 \frac{A}{cm^2}$ reversal.

In fig. 4.39 the current induced switching at low temperature is shown. The decrease in critical current from low $2\times10^7 \frac{A}{cm^2}$ to low $1\times10^7 \frac{A}{cm^2}$, namely $1.15\times10^7 \frac{A}{cm^2}$ for AP-P transition and $1.48\times10^7 \frac{A}{cm^2}$ for P-AP transition is in the right order for applications with industrial interest. One of the reasons for such low values is the higher spin polarization and the low damping for both used materials. The damping of the permalloy is 6×10^{-3} [Koba 09] whilst the damping for the used NiMnSb is in the low 10^{-3}-range. A remarkable feature in both, the room temperature and the low temperature, curves is the increase in resistance for the antiparallel configuration when decreasing the current. This increase can be seen as an indication for an intermediate configuration state with the ferromagnetic layers being not completely antiparallel aligned. Comparing the magnetoresistance curves

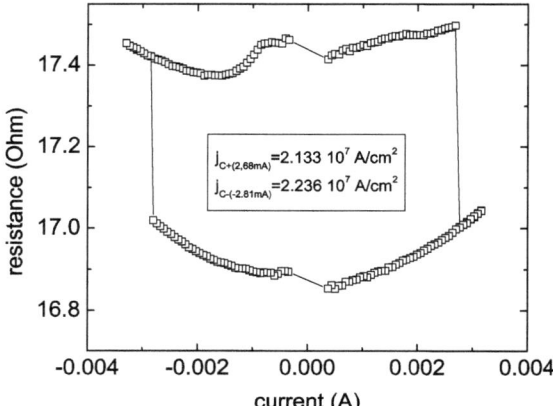

Fig. 4.38: Room temperature current induced switching measurement of a elliptical shaped (80 × 200 nm²) NiMnSb based pseudo-spin-valve with the same composition ((40)NiMnSb/(10)Cu/(6)Py) as above

which show no significant slope in the field dependence neither for the antiparallel state nor for the parallel state, the configuration at around zero amperes can be seen as the 'most antiparallel state'. The decrease of the resistance with increasing absolute current might be seen as the reason for the later reported oscillations. Another feature in the reported curve is the decrease of the resistance between 1.8 mA and 2.2 mA for the room temperature measurement and the step in the curve at 1.1 mA for low temperature. One explanation for such changing can be the presence of a multi-domain state which is suppressed by the field but can be established in the absence of an external field. The smooth changing for higher temperature in combination with the intermediate changing for lower temperature is strengthen the explanation.

4.6.3 spin-torque oscillator

Preliminary measurements have always been performed before starting to measure the emitted high frequency power. The current induced switching and the magnetoresistance have been measured to ensure the pillars are working and exhibiting comparable characteristics as the DC structures. The fig. 4.40 shows the magnetoresistance measurement of the STO device. Black line is the up-sweep (starting from high negative fields to ensure parallel alignment) exhibiting a clear switching after zero field and the red curve is the

4.6. Spin-torque devices using NiMnSb

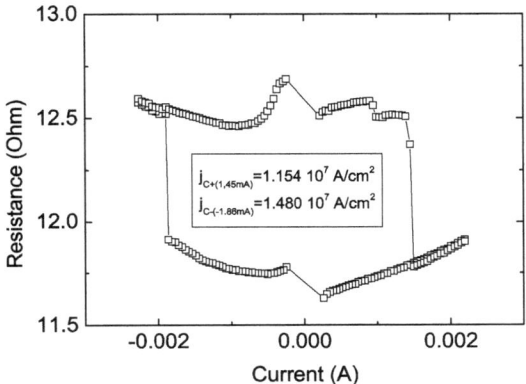

Fig. 4.39: Low temperature measurement of the same device presented above offers an expected decrease in current densities which are necessary to change the magnetic configuration

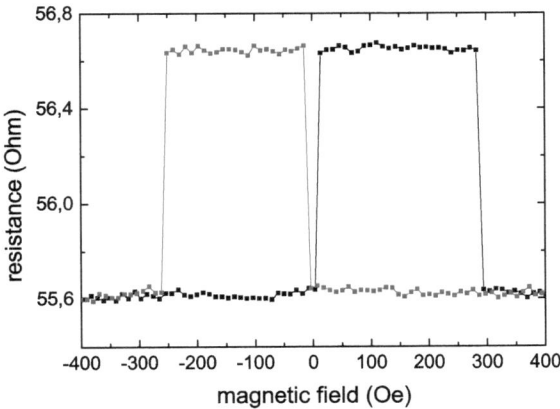

Fig. 4.40: Preliminary magnetoresistance measurement of the STO device (elliptical shaped $(100 \times 200 \text{ nm}^2$; (40)NiMnSb/(10)Cu/(6)Py) to verify the DC-characteristics

down sweep.

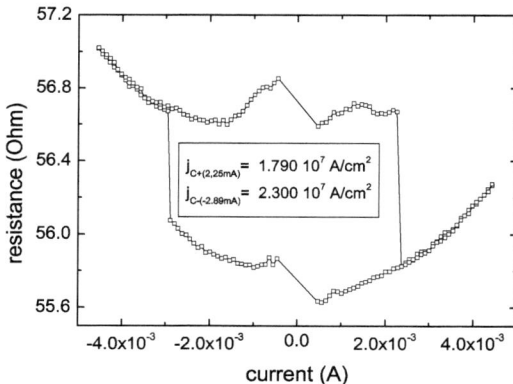

Fig. 4.41: Resistance versus current is plotted. Two well separated states are visible(high resistance, antiparallel alignment, low resistance, parallel alignment) and the corresponding current densities of $1.79 \times 10^7 A/cm^2$ for switching from anti- to parallel and $2.3 \times 10^7 A/cm^2$ to reverse the configuration.

The comparable low current densities are indicating the same quality as for the DC devices. A small asymmetry in the switching curve can be understood by taking the higher remanence field of the magnet at the high frequency setup compared to the magnet of the DC probe-station into account. The resistance change from 17 Ω for the DC- to 56 Ω for the AC-devices is artificial, and originates in a change of the detecting method from '3-point'- to '2-point'-measurements.

The contact layout used to measure the high frequency emitting spin-torque oscillators is shown in the Annex (see fig. A.2.j). Using a coplanar waveguide geometry with the nano-pillar contacting the signal line and the backside contact shorted to ground line of the waveguide offers best impedance matching to the tips in ground-signal-ground (gsg) geometry. Nevertheless the impedance mismatch between the waveguide and the pillar causes part of the emitted power to be reflected before entering the CPW. This results in a lowered measured output power. Fig. 4.42 shows the electrical circuit to measure the spin-torque oscillators. A standard analogue-digital output card for personal computers is used to apply a voltage. The biased current is measured using a reference resistor of 10 Ω connected in series. Additionally a bias-T is used to split and combine the alternating high frequency signal emitted by the pillar and the constant applied voltage. To amplify the signal a broad band amplifier with a gain of 34 dB is used. The measured power of the spectrum at the analyser can be recalculated to the emitted power of the device by

4.6. Spin-torque devices using NiMnSb

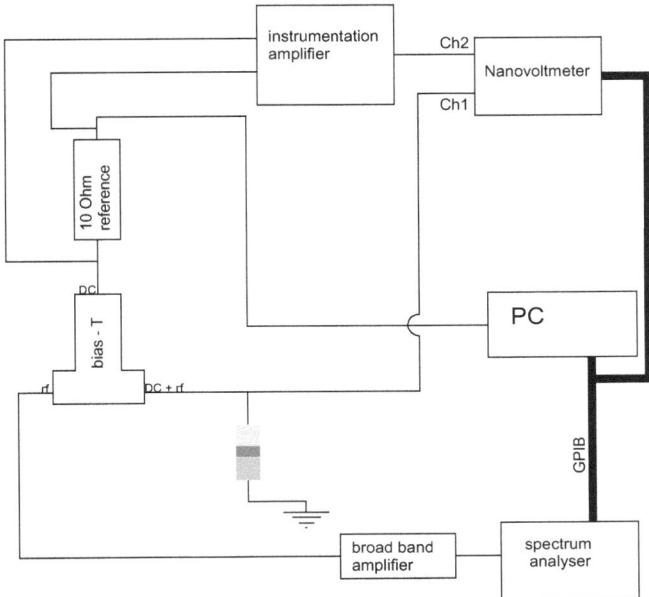

Fig. 4.42: Schematics of the measurement of spin-torque oscillators neglecting the impedance mismatch between pillar and waveguide.

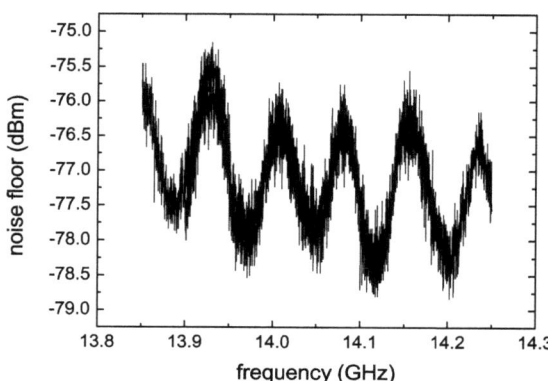

Fig. 4.43: Noise figure of the measurement setup, measured with the tips lowered on the sample but without an applied current

Fig. 4.43 shows the background or the noise floor. The spectrum was measured with the sample contacted but without applying a current and therefore without any effects from the sample. An origin of the waviness of the signal interacting with the mixer in the spectrum analyser. This reference spectrum was subtracted for each measurements in order to have power-of-signal over noise floor. Recalculating the real noise floor one has to subtract from the signal in fig. 4.43 the amplification of the used broad band amplifier, namely 34 dB. This results in a noise floor of the measurement of -134 dBm. By using the following equation one can show that this is the normal thermal background [Poza 05].

$$P_n = kTB \tag{4.5}$$

In equation 4.5 k is the Boltzmann constant, T the temperature and B the band-width. Taking T = 290 K for room temperature and the bandwidth as B = 100 MHz the calculated noise power is -134 dB.

All results below are measured by establishing a parallel alignment using a high negative magnetic field and then sweeping to positive fields, therefore changing the magnetic configuration to anti-parallel. Fig. 4.44 shows the history of measurements.

After establishing the anti-parallel configuration a negative current is applied.

4.6. Spin-torque devices using NiMnSb

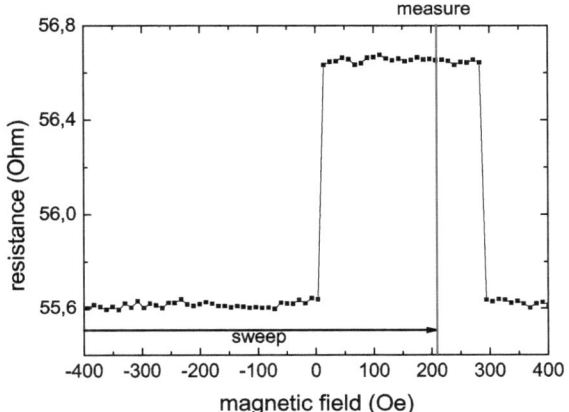

Fig. 4.44: Illustrated procedure to prepare the device in anti-parallel magnetic configuration before applying current and measure spin-torque oscillations

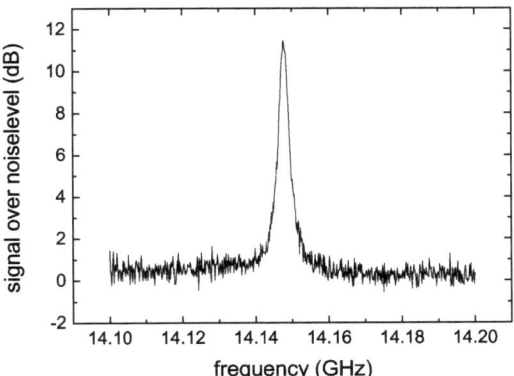

Fig. 4.45: Spectrum for the pseudo-spin-valve ((40)NiMnSb/(10)Cu/(6)Py) with an applied current of -2.983 mA

The figure 4.45 shows the emitted high frequency when biasing the pillar with a negative current of -2.983 mA. The shown frequency is emitted with an external field of 207

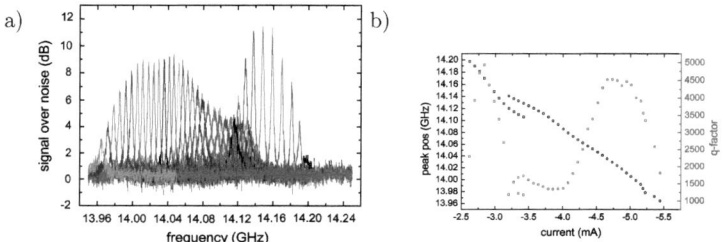

Fig. 4.46: Spectrum for the device measured in an external field of 207 Oe for different currents (a) and the corresponding peak position and q-factors over applied currents (b)

Oe applied. By varying the current the frequency shifts due to the current dependent spin torque.

In fig. 4.46.a) the spectrum for different currents is shown. With decreasing current the frequency shifts to lower values (red-shift). This has been observed in the literature and attributed to in-plane precession [Grol 06]. For an intermediate regime, namely between -3.1 mA and -3.5 mA, two peaks become visible (see 4.47).

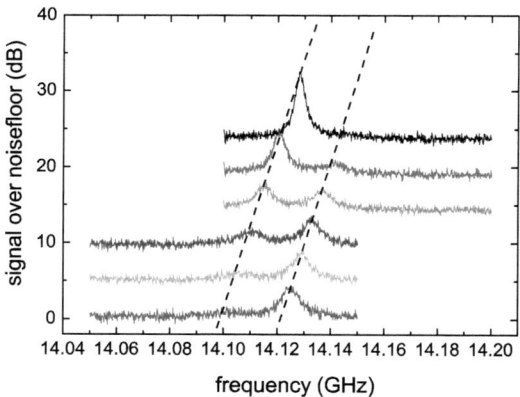

Fig. 4.47: Spectrum for different currents (black:-3.128 mA; red:-3.2021 mA; green:-3.2759 mA; blue:-3.3452 mA; cyan:-3.4207 mA; magenta:-3.4935 mA)

Fig. 4.47 shows the spectra for the currents between -3.128 mA and -3.4935 mA. In this spectral part two well separated peaks appear. This can be explained as different modes which manifest inside the spin-valve. Another evidence for two separate modes can be seen in the q-factor over current plot as shown in 4.46.b).

4.6. Spin-torque devices using NiMnSb

The emitted frequency, plotted on the left y-axis, and the corresponding q-factor, plotted on the right y-axis, is shown versus the applied current. In a confined current range the two frequencies of the peaks can be identified. On the other hand the q-factor shows a texture with different maximums and minimums. The first maximum corresponds to the first established mode. Values for the q-factor between 1000 and 8000 are cited for devices in an external field [Devo 09, Bone 09].Most of these high q-factors were measured in external fields in the order of Tesla whilst the presented device is operating in a low field regime.

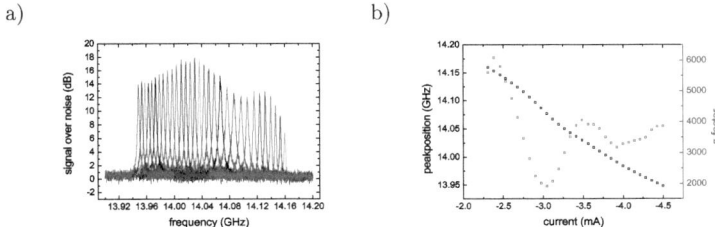

Fig. 4.48: Measurement of the spectrum of the same device in an external applied magnetic field of 169 Oe (a) and the current dependencies for q-factor and peak position is shown in b)

A reduced operating field of 169 Oe results in a lower emitted frequencies whilst having the same systematical behaviour. The frequency shows a red shift with decreasing current as measured in a field of 207 Oe. Fig. 4.48.b) shows the peak position and the q-factor of the system. In the texture of the q-factor not as many maximums and minimums are visible.

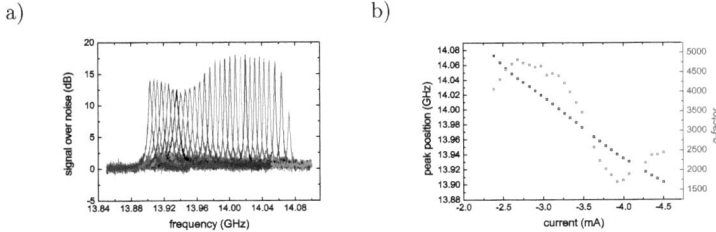

Fig. 4.49: Spectrum for the oscillator in an external field of 135 Oe (a) and in b) the corresponding current dependent peak position and q-factors are shown

Further decrease in field shifts the frequency range to lower values whilst offering the same characteristic red-shift of the emitted frequency dependent of the current. As well for the q-factors. With the used procedure of preparing an anti-parallel state it is also possible to operate the oscillator in the absence of an external field. The fig. 4.50 shows the peak position of the maximum of the first (square) and the second (triangle) 'mode' of the device.

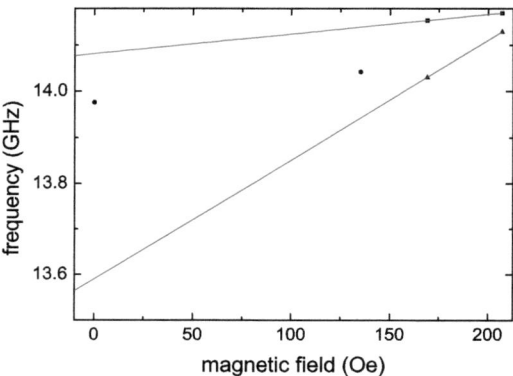

Fig. 4.50: Peak position of the first modes of the device over the applied external field is shown, higher fields (207 and 169 Oe) show two modes while the spectrum for lower fields (135 and 0 Oe) show only one mode which is likely a distinct mode due to the fact that they show not the same field dependency as the measured modes for higher fields

The slope of the extrapolation to zero field between the first and the second 'mode' is different and therefore results in a zero field splitting. The circles are the emitted frequencies of the first mode of the device in an external field of 135 Oe and without external field respectively. Both frequencies do not lie on the extended straight line of the two modes for higher fields and are therefore different modes. The presence of an oscillation in the absence of an external field is one of the main goals for industrial use.

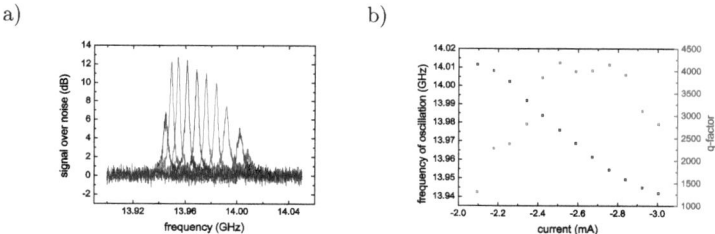

Fig. 4.51: Emitted high frequency of the device operating in the absence of an applied external magnetic field (a) and the frequency and q-factor over applied currents (b) are shown

4.6. Spin-torque devices using NiMnSb

As shown in fig. 4.51 a) and b) the auto-oscillations at zero field can cover a bandwidth of 50 MHz with an applied current between -2.1 mA to -3.0 mA. Therefore the ratio of frequency shift to current is $\frac{50 \text{ MHz}}{1.9 \text{ mA}} = 26.32 \frac{\text{MHz}}{\text{mA}}$. The outstanding high q-factor of 4180 for a spin-torque oscillator working in the absence of an external field is one order of magnitude higher than the reported value of 280 by [Devo 07]. The reason for the unexpected high emitted frequency in combination with the purity of the signal is not yet understood. One explanation can be a coupled system of both ferromagnetic layers. This can also explain the presence of the oscillation for negative currents which favours the system to be anti-parallel (see current induced switching measurement 4.4). An increase in current above the threshold current densities necessary to switch the magnetic configuration causes the 'fixed' layer to precess as well. The frequency range in combination with the q-factor and the DC characteristics of the presented pseudo-spin-valve with NiMnSb and Py makes it very interesting for industrial use.

Chapter 5
Conclusion and outlook

The outstanding low damping for NiMnSb makes this half-Heusler alloy a promising candidate for industrial use. Whilst preliminary reports have already given hints on its good high frequency characteristics for thin NiMnSb layers the development in growth of the material can meanwhile ensure a remaining good damping for high thicker crystalline layers. A direct result of the low damping was the measurement of magnetostatic modes in sub-micrometer sized elements. The modes could be observed using a sample with only a few stripes which is equivalent to a decreased volume compared to reported magneto static modes observations.

According to theory the half-Heusler alloy should show a 100% spin polarization. Measurements of the magnetoresistance using NiMnSb in combination with Permalloy did not show this value but a ratio of 3.4 % at room temperature and almost double at lower temperature. The current densities necessary to switch the magnetic formation is in the low 10^7 A/cm^2 which is in the comparable to state of the art metal based devices. The most outstanding results presented in this thesis are the measurements of the spin-torque oscillations of NiMnSb based pseudo-spin-valves. By the use of the half-Heusler alloy spin-torque oscillators could be realized which operates in the absence of an external magnetic field. These oscillators can compete state of the art devices in most of the specifications whilst offering a pure signal with an outstanding high q-factor. The q-factor is almost an order of magnitude higher then the best reported value for a spin-torque oscillator which is working without an external field. One of the main goals for the future devices will be to improve the emitted output power. The measured power of the devices presented in the thesis is in the order of 300 pW. This is not a problem only for NiMnSb based devices but for all devices in the field of spin-torque oscillators. The authors Mancoff et al. [Manc 05] presented promising measurements of locked STO with an increased output power. Recently promising experiments of coupled nano-point contacts [Ruot 09] have also been reported and therefore the right way to increase the power is drawn. By optimizing geometrical parameters and paying attention to the anisotropy fields in the NiMnSb it will be possible to build such devices consisting of two or more spin-torque oscillators whilst operating in the absence of an external field.

Appendix A

Appendices

The following appendix contains additional background to scattering processes and the sample fabrication based on the Spintronic lecture of R. Gross and A. Marx at the Walther-Meisner Institut in Garching.

A.1 Scattering

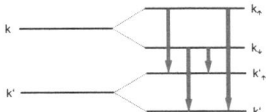

Fig. A.1: schematics of the spin dependent scattering

A scattering process in metals transfers electrons with the energy E_k from state $\Psi(\mathbf{r}, E_k)$ in the state $\Psi(\mathbf{r}, E_{k'})$ with energy $E_{k'}$. The scattering is described with the matrix

$$T_{kk'} = \frac{1}{V} \int d^3\mathbf{r}\, \psi_k(\mathbf{r}) \Delta V^\sigma(\mathbf{r}) \psi_{-k'}(\mathbf{r}) \tag{A.1}$$

Hereby is σ either $\uparrow$ or $\downarrow$ and the ΔV^σ describes a spin dependent scattering potential. The transition probability can be written:

$$P_{kk'} = \frac{2\pi}{\hbar} n_d D(E) |T_{kk'}^2| \delta(E_k - E_{k'}) \tag{A.2}$$

The delta function is used for elastically scattering. The n_d is the concentration of defects. As long as the concentration of the defect can be regarded as separated (diluted limit) the probability is proportional to n_d and the density of states $D(E)$. For the relevant energy we can assume $D(E)$ to be equal to $D(E_F)$. In ferromagnetic materials the transition matrix is:

$$P_{kk'} = \begin{pmatrix} P_{kk'}^{\uparrow\uparrow} & P_{kk'}^{\uparrow\downarrow} \\ P_{kk'}^{\downarrow\uparrow} & P_{kk'}^{\downarrow\downarrow} \end{pmatrix} \tag{A.3}$$

This elements correspond to two spin obtaining processes and two spin-flip processes. Figure A.1 illustrates the transitions. Summing up all possible states in which the electron can be scattered one gets the inverse of the relaxation time.

$$\tau_k^{-1} = \sum_{k'} p_{kk'}^{\sigma} \tag{A.4}$$

The relaxation time is in general spin and state dependent. To simplify matters the state dependency is averaged over the Fermi-surface.

$$\langle \tau_k \rangle = \frac{\sum_k \delta(E_k - E_F)\tau_k}{\sum_k \delta(E_k - E_F)} \tag{A.5}$$

The spin dependent scattering differs for $\uparrow$ and $\downarrow$ due to the potentials and the electrical structure which gives a difference in the relaxation times. This can be expressed with the spin-anisotropy.

$$\beta = \frac{\tau^{\uparrow}}{\tau^{\downarrow}} \tag{A.6}$$

According to equation A.6 for $\beta < 1$ the scattering of the majority spins is dominant and for $\beta > 1$ the minority spin scattering is more dominant. Using the two spin channel model to describe the GMR effect this is only valid if the spin flip process do not mix the channels for $\uparrow$ and $\downarrow$.

One reason for scattering can be a defect in the ferromagnet. The interaction of the conducting electron with spin **s** with the conducting electron of the defect with spin S_i at the place R_i can be described with the exchange integral

$$V_S = -\frac{J}{n_D} \sum_i \mathbf{s} \cdot \mathbf{S}_i \delta(\mathbf{r} - \mathbf{R}_i) = -\frac{J}{n_D} \sum_i (\sigma^x S_i^x + \sigma^y S_i^y + \sigma^z S_i^z) \delta(\mathbf{r} - \mathbf{R}_i) \tag{A.7}$$

With $\sigma^i, i = x, y, z$ are the Pauli spin matrices.

$$\sigma^x = \frac{\hbar}{2} \begin{pmatrix} 0 & 1 \\ 1 & 0 \end{pmatrix} \quad \sigma^y = \frac{\hbar}{2} \begin{pmatrix} 0 & -i \\ i & 0 \end{pmatrix} \quad \sigma^z = \frac{\hbar}{2} \begin{pmatrix} 1 & 0 \\ 0 & -1 \end{pmatrix} \tag{A.8}$$

and $n_D = N_D/V$ is the density of defects. The exchange constant J depends on the gap between the conducting electrons and the defect center.

A.2 Sample fabrication

In the used MBE cluster of the department of Experimentelle Physik III it is possible to transfer the NiMnSb layers grown in the Heusler-chamber directly into the sputtering chamber without breaking the vacuum. This is important to avoid oxidation on the surface of the NiMnSb layer and consequently the CuO interlayer. The background pressure in the MBE cluster is in the order of 10^{-10} and in the sputtering chamber 10^{-9}. After sputtering the heterostructure the sample is transfered out and processed.

Fig A.2 a) to i) show the process of fabricating samples for DC-measurements whilst for AC-measurements the main process remains the same but the top contact as shown in j). After spinning an e-Beam resist (thickness approximately 600 nm) on top of the heterostructure the pillars with smallest dimension of 80x200 nm^2 (A.2 are exposed using a secondary electron microscope (SEM) in combination of a pattern unit (fig. A.2b). The resist is used negative by a reversal backing and an additional flood-expose using ultraviolet light. In fig. A.2c) the developed resist is used as a etching mask. By the use of ion milling with argon in the chemical assisted ion beam etch machine (CAIBE) the pillars are etched down to the NiMnSb layer which is used as backside contact. The etched pillars are transfered immediately in the plasma enhanced vapor deposition machine (PECVD) in which an isolating layer Si_3N_4 with a thickness of 20 to 30 nm is deposited (A.2.d). The isolating layer is covering the sample with the pillars and the resist as well. In the next step the acetone lift-off is done using tempered acetone and ultrasonic. Consequently the resist under the isolator is removed leaving a hole in the Si_3N_4 and allows to contact the top of the pillars (A.2.e). The backside contacts are defined using an electron resist and exposing the resist by e-beam. Afterwards the developed resist is used as an etch mask for opening the backside contacts using reactive ion etching (RIE). After removing the resist using acetone the top-contacts and the backside-contacts are open to contact with pads (see fig.A.2.g). Two different layouts are used, for DC characterization and for ac characterization (A.2.i and j). The geometry used for DC measurements allows a quasi 4-point measurement (see 4.6.1 for detailed description). While the AC layout allows only 2-point measurements but, due to the geometry of the signal line and the ground in coplanar waveguide geometry [Rieg 06], measurements in the GHz range.

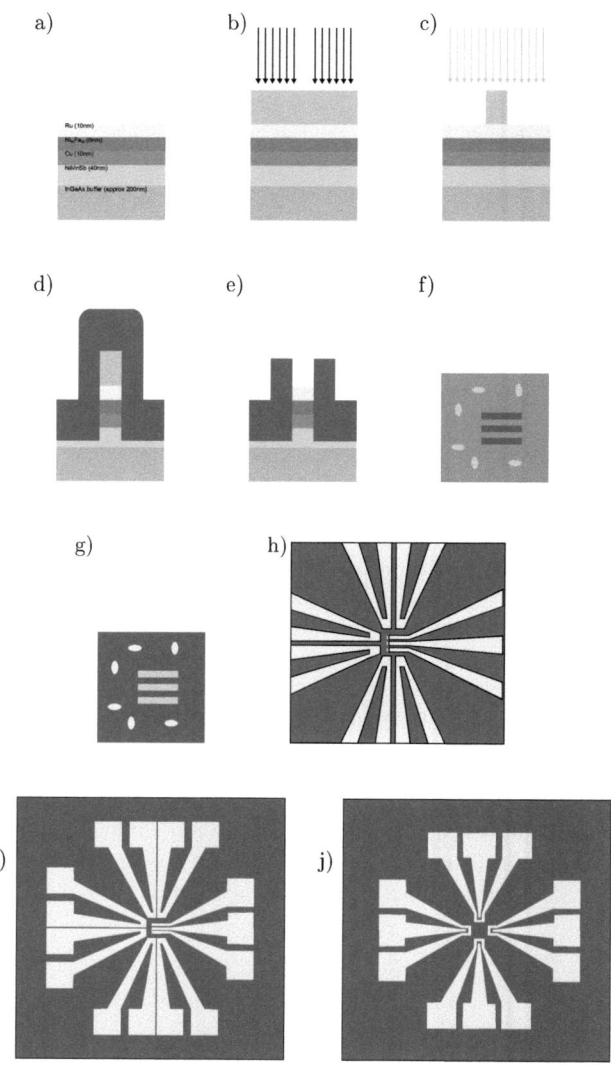

Fig. A.2: process starts from the unpatterned heterostructure (a) to the final device for DC characterization (i) and for AC measurements (j)

Bibliography

[Abri 62]   A. A. Abrikosos and L. P. Gorkov. "Spin-orbit Interaction and the Knight Shift In Superconductors". *Soviet Physics Jetp-ussr*, Vol. 15, No. 4, pp. 752–757, 1962.

[Aria 99]   R. Arias and D. L. Mills. "Extrinsic contributions to the ferromagnetic resonance response of ultrathin films". *Physical Review B*, Vol. 60, No. 10, pp. 7395–7409, Sep. 1999.

[Baib 88]   M. N. Baibich, J. M. Broto, A. Fert, F. N. Vandau, F. Petroff, P. Eitenne, G. Creuzet, A. Friederich, and J. Chazelas. "Giant Magnetoresistance of (001)fe/(001) Cr Magnetic Superlattices". *Physical Review Letters*, Vol. 61, No. 21, pp. 2472–2475, Nov. 1988.

[Bail 06]   M. Bailleul, R. Hollinger, and C. Fermon. "Microwave spectrum of square Permalloy dots: Quasisaturated state". *Physical Review B*, Vol. 73, No. 10, p. 104424, March 2006.

[Berg 96]   L. Berger. "Emission of spin waves by a magnetic multilayer traversed by a current". *Physical Review B*, Vol. 54, No. 13, pp. 9353–9358, Oct. 1996.

[Bina 89]   G. Binasch, P. Grunberg, F. Saurenbach, and W. Zinn. "Enhanced Magnetoresistance In Layered Magnetic-structures With Antiferromagnetic Interlayer Exchange". *Physical Review B*, Vol. 39, No. 7, pp. 4828–4830, March 1989.

[Bona 85]   G. L. Bona, F. Meier, M. Taborelli, E. Bucher, and P. H. Schmidt. "Spin Polarized Photoemission From Nimnsb". *Solid State Communications*, Vol. 56, No. 4, pp. 391–394, 1985.

[Bone 09]   S. Bonetti, P. Muduli, F. Mancoff, and J. Akerman. "Spin torque oscillator frequency versus magnetic field angle: The prospect of operation beyond 65 GHz". *Applied Physics Letters*, Vol. 94, No. 10, p. 102507, March 2009.

[Borc 01]   C. N. Borca, T. Komesu, H. K. Jeong, P. A. Dowben, D. Ristoiu, C. Hordequin, J. P. Nozieres, J. Pierre, S. Stadler, and Y. U. Idzerda. "Evidence for temperature dependent moments ordering in ferromagnetic NiMnSb(100)". *Physical Review B*, Vol. 64, No. 5, p. 052409, Aug. 2001.

[Brya 93] P. H. Bryant, J. F. Smyth, S. Schultz, and D. R. Fredkin. "Magnetostatic-mode spectrum of rectangular ferromagnetic particles". *Phys. Rev. B*, Vol. 47, No. 17, pp. 11255–, May 1993.

[Caba 98] J. A. Caballero, Y. D. Park, J. R. Childress, J. Bass, W. C. Chiang, A. C. Reilly, W. P. Pratt, and F. Petroff. "Magnetoresistance of NiMnSb-based multilayers and spin valves". May 1998.

[DeGr 83] R. A. DeGroot, F. M. Mueller, P. G. Vanengen, and K. H. J. Buschow. "New Class Of Materials - Half-Metallic Ferromagnets". *Physical Review Letters*, Vol. 50, No. 25, pp. 2024–2027, 1983.

[Devo 07] T. Devolder, A. Meftah, K. Ito, J. A. Katine, P. Crozat, and C. Chappert. "Spin transfer oscillators emitting microwave in zero applied magnetic field". *Journal Of Applied Physics*, Vol. 101, No. 6, p. 063916, March 2007.

[Devo 09] T. Devolder, L. Bianchini, J. V. Kim, P. Crozat, C. Chappert, S. Cornelissen, M. O. de Beeck, and L. Lagae. "Auto-oscillation and narrow spectral lines in spin-torque oscillators based on MgO magnetic tunnel junctions". *Journal of Applied Physics*, Vol. 106, No. 10, p. 103921, Nov. 2009.

[Dona 99] M. Donahue and D. Porter. *OOMMF User's Guide, Version 1.0*. National Institute of Standards and Technology, Gaithersburg, MD (Sept 1999), interagency report nistir 6376 Ed., 1999.

[Ever 98] B. A. Everitt and A. V. Pohm. "Pseudo spin valve magnetoresistive random access memory". *Journal of Vacuum Science & Technology A-vacuum Surfaces and Films*, Vol. 16, No. 3, pp. 1794–1800, May 1998.

[Fert 04] A. Fert, V. Cros, J. M. George, J. Grollier, H. Jaffres, A. Hamzic, A. Vaures, G. Faini, J. Ben Youssef, and H. Le Gall. "Magnetization reversal by injection and transfer of spin: experiments and theory". *Journal of Magnetism and Magnetic Materials*, Vol. 272, pp. 1706–1711, May 2004.

[Gall 97] W. J. Gallagher, S. S. P. Parkin, Y. Lu, X. P. Bian, A. Marley, K. P. Roche, R. A. Altman, S. A. Rishton, C. Jahnes, T. M. Shaw, and G. Xiao. "Microstructured magnetic tunnel junctions". *Journal of Applied Physics*, Vol. 81, No. 8, pp. 3741–3746, Apr. 1997.

[Grol 06] J. Grollier, V. Cros, and A. Fert. "Synchronization of spin-transfer oscillators driven by stimulated microwave currents". *Physical Review B*, Vol. 73, No. 6, p. 060409, Feb. 2006.

[Hein 04] B. Heinrich, G. Woltersdorf, R. Urban, O. Mosendz, G. Schmidt, P. Bach, L. Molenkamp, and E. Rozenberg. "Magnetic properties of NiMnSb(001) films grown on InGaAs/InP(001)". June 2004.

[Hein 67] B. Heinrich, D. Fraitova, and Kambersk.V. "Influence of S-d Exchange On Relaxation of Magnons In Metals". *Physica Status Solidi*, Vol. 23, No. 2, pp. 501–&, 1967.

[Heus 03] F. Heusler. *Verh. Deutsche Physikalische Gesellschaft*, Vol. 5, p. 219 ff., 1903.

[Hord 98] C. Hordequin, J. P. Nozieres, and J. Pierre. "Half metallic NiMnSb-based spin-valve structures". *Journal of Magnetism and Magnetic Materials*, Vol. 183, No. 1-2, pp. 225–231, March 1998.

[Huai 04] Y. M. Huai, F. Albert, P. Nguyen, M. Pakala, and T. Valet. "Observation of spin-transfer switching in deep submicron-sized and low-resistance magnetic tunnel junctions". *Applied Physics Letters*, Vol. 84, No. 16, pp. 3118–3120, Apr. 2004.

[Hurb 98] M. J. Hurben and C. E. Patton. "Theory of two magnon scattering microwave relaxation and ferromagnetic resonance linewidth in magnetic thin films". *Journal of Applied Physics*, Vol. 83, No. 8, pp. 4344–4365, Apr. 1998.

[Jede 01] F. J. Jedema, A. T. Filip, and B. J. van Wees. "Electrical spin injection and accumulation at room temperature in an all-metal mesoscopic spin valve". *Nature*, Vol. 410, No. 6826, pp. 345–348, March 2001.

[Jorz 99] J. Jorzick, S. O. Demokritov, C. Mathieu, B. Hillebrands, B. Bartenlian, C. Chappert, F. Rousseaux, and A. N. Slavin. "Brillouin light scattering from quantized spin waves in micron-size magnetic wires". *Physical Review B*, Vol. 60, No. 22, pp. 15194–15200, Dec. 1999.

[Kala 06] S. S. Kalarickal, P. Krivosik, M. Z. Wu, C. E. Patton, M. L. Schneider, P. Kabos, T. J. Silva, and J. P. Nibarger. "Ferromagnetic resonance linewidth in metallic thin films: Comparison of measurement methods". *Journal of Applied Physics*, Vol. 99, No. 9, p. 093909, May 2006.

[Kati 00] J. A. Katine, F. J. Albert, R. A. Buhrman, E. B. Myers, and D. C. Ralph. "Current-driven magnetization reversal and spin-wave excitations in Co/Cu/Co pillars". *Physical Review Letters*, Vol. 84, No. 14, pp. 3149–3152, Apr. 2000.

[Kitt 49] C. Kittel. "Physical Theory Of Ferromagnetic Domains". *Reviews Of Modern Physics*, Vol. 21, No. 4, pp. 541–583, 1949.

[Koba 09] K. Kobayashi, N. Inaba, N. Fujita, Y. Sudo, T. Tanaka, M. Ohtake, M. Futamoto, and F. Kirino. "Damping Constants for Permalloy Single-Crystal Thin Films". *Ieee Transactions On Magnetics*, Vol. 45, No. 6, pp. 2541–2544, June 2009.

[Kove 05] A. Koveshnikov, G. Woltersdorf, J. Q. Liu, B. Kardasz, O. Mosendz, B. Heinrich, K. L. Kavanagh, P. Bach, A. S. Bader, C. Schumacher, C. Ruster, C. Gould, G. Schmidt, L. W. Molenkamp, and C. Kumpf. "Structural and magnetic properties of NiMnSb/InGaAs/InP(001)". *Journal Of Applied Physics*, Vol. 97, No. 7, p. 073906, Apr. 2005.

[Land 08] P. Landeros, R. E. Arias, and D. L. Mills. "Two magnon scattering in ultrathin ferromagnets: The case where the magnetization is out of plane". *Physical Review B*, Vol. 77, No. 21, p. 214405, June 2008.

[Land 35] L. D. Landau and E. Lifshitz. "On the theory of the dispersion of magnetic permeability in ferromagnetic bodies". *Physical Z. Sowietunion*, Vol. 8, p. 153, 1935.

[Li 09] X. A. Li, Z. Z. Zhang, Q. Y. Jin, and Y. W. Liu. "Spin-torque-induced switching in a perpendicular GMR nanopillar with a soft core inside the free layer". *New Journal of Physics*, Vol. 11, p. 023027, Feb. 2009.

[Loch 10a] F. Lochner, A. Riegler, G. Schmidt, and L. Molenkamp. "High magnetic and crystalline NiMnSb layers". *yet unpublished*, No. xxxxxx, 2010.

[Loch 10b] F. Lochner. *Epitaxial growth and characterization of NiMnSb layers for novel spintronic devices*. PhD thesis, 2010.

[Loub 09] G. de Loubens, A. Riegler, B. Pigeau, F. Lochner, F. Boust, K. Y. Guslienko, H. Hurdequint, L. W. Molenkamp, G. Schmidt, A. N. Slavin, V. S. Tiberkevich, N. Vukadinovic, and O. Klein. "Bistability of Vortex Core Dynamics in a Single Perpendicularly Magnetized Nanodisk". *Physical Review Letters*, Vol. 102, No. 17, p. 177602, May 2009.

[Manc 05] F. B. Mancoff, N. D. Rizzo, B. N. Engel, and S. Tehrani. "Phase-locking in double-point-contact spin-transfer devices". *Nature*, Vol. 437, No. 7057, pp. 393–395, Sep. 2005.

[Mang 06] S. Mangin, D. Ravelosona, J. A. Katine, M. J. Carey, B. D. Terris, and E. E. Fullerton. "Current-induced magnetization reversal in nanopillars with perpendicular anisotropy". *Nature Materials*, Vol. 5, No. 3, pp. 210–215, March 2006.

[Mede 78] R. Medervey and P. M. Tedrow. "Surface Relaxation-times of Conduction-electron Spins In Superconductors and Normal Metals". *Physical Review Letters*, Vol. 41, No. 12, pp. 805–808, 1978.

[Mizu 01a] S. Mizukami, Y. Ando, and T. Miyazaki. "Ferromagnetic resonance linewidth for NM/80NiFe/NM films (NM = Cu, Ta, Pd and Pt)". *Journal of Magnetism and Magnetic Materials*, Vol. 226, pp. 1640–1642, May 2001.

[Mizu 01b] S. Mizukami, Y. Ando, and T. Miyazaki. "The study on ferromagnetic resonance linewidth for NM/80NiFe/NM (NM = Cu, Ta, Pd and Pt) films". *Japanese Journal of Applied Physics Part 1-regular Papers Short Notes & Review Papers*, Vol. 40, No. 2A, pp. 580–585, Feb. 2001.

[Myer 99] E. B. Myers, D. C. Ralph, J. A. Katine, R. N. Louie, and R. A. Buhrman. "Current-induced switching of domains in magnetic multilayer devices". *Science*, Vol. 285, No. 5429, pp. 867–870, Aug. 1999.

[Nogu 99] J. Nogues and I. K. Schuller. "Exchange bias". *Journal of Magnetism and Magnetic Materials*, Vol. 192, No. 2, pp. 203–232, Feb. 1999.

[Otto 89] M. J. Otto, R. A. M. VanWoerden, P. J. VanDervalk, J. Wijngaard, C. F. VanBruggen, C. Haas, and K. H. J. Buschow. "Half-metallic Ferromagnets .1. Structure and Magnetic-properties of Nimnsb and Related Inter-metallic Compounds". *Journal of Physics-condensed Matter*, Vol. 1, No. 13, pp. 2341–2350, Apr. 1989.

[Pige 10a] B. Pigeau, G. de Loubens, O. Klein, A. Riegler, F. Lochner, G. Schmidt, L. W. Molenkamp, V. S. Tiberkevich, and A. N. Slavin. "A frequency-controlled magnetic vortex memory". *Applied Physics Letters*, Vol. 96, No. 13, p. 132506, March 2010.

[Pige 10b] B. Pigeau, G. de Loubens, O. Klein, A. Riegler, F. Lochner, G. Schmidt, and L. W. Molenkamp. "Optimal control of vortex core polarity by resonance microwave pulses". *Nature*, Vol. DOI: 10.1038/nphys1810, 2010.

[Poza 05] D. M. Pozar. *Microwave Engineering*. John Willey & Sons, Inc., 2005.

[Rave 06] D. Ravelosona, S. Mangin, Y. Lemaho, J. A. Katine, B. D. Terris, and E. E. Fullerton. "Domain wall creation in nanostructures driven by a spin-polarized current". *Physical Review Letters*, Vol. 96, No. 18, p. 186604, May 2006.

[Rieg 06] A. Riegler. *time-domain Messungen der ferromagnetischen Resonanz an NiMnSb Schichten*. Master's thesis, University of Wuerzburg, 2006.

[Rieg 10a] A. Riegler, F. Lochner, C. Gould, G. Schmidt, and L. Molenkamp. "Magnetostatic Modes in Arrays of Sub-Micrometer Sized Elements of (001)NiMnSb". *yet unpublished*, 2010.

[Rieg 10b] A. Riegler, F. Lochner, C. Gould, G. Schmidt, and L. Molenkamp. "Substrate Induced Uniaxial Anisotropies in (111)NiMnSb Layers". *yet unpublished*, 2010.

[Rieg 10c] A. Riegler, F. Lochner, C. Gould, G. Schmidt, and L. Molenkamp. "Very High Q-Factors for Spin-Torque Oscillators in Zero Field". *yet unpublished*, 2010.

[Rous 01] Y. Roussigne, S. M. Cherif, C. Dugautier, and P. Moch. "Experimental and theoretical study of quantized spin-wave modes in micrometer-size permalloy wires". *Physical Review B*, Vol. 63, No. 13, p. 134429, Apr. 2001.

[Ruot 09] A. Ruotolo, V. Cros, B. Georges, A. Dussaux, J. Grollier, C. Deranlot, R. Guillemet, K. Bouzehouane, S. Fusil, and A. Fert. "Phase-locking of magnetic vortices mediated by antivortices". *Nature Nanotechnology*, Vol. 4, No. 8, pp. 528–532, Aug. 2009.

[Slon 96]   J. C. Slonczewski. "Current-driven excitation of magnetic multilayers". *Journal Of Magnetism And Magnetic Materials*, Vol. 159, No. 1-2, pp. L1–L7, June 1996.

[Soul 98]   R. J. Soulen, J. M. Byers, M. S. Osofsky, B. Nadgorny, T. Ambrose, S. F. Cheng, P. R. Broussard, C. T. Tanaka, J. Nowak, J. S. Moodera, A. Barry, and J. M. D. Coey. "Measuring the spin polarization of a metal with a superconducting point contact". *Science*, Vol. 282, No. 5386, pp. 85–88, Oct. 1998.

[Sun 99]   J. Z. Sun. "Current-driven magnetic switching in manganite trilayer junctions". *Journal of Magnetism and Magnetic Materials*, Vol. 202, No. 1, pp. 157–162, July 1999.

[Tana 99]   C. T. Tanaka, J. Nowak, and J. S. Moodera. "Spin-polarized tunneling in a half-metallic ferromagnet". *Journal of Applied Physics*, Vol. 86, No. 11, pp. 6239–6242, Dec. 1999.

[Tehr 99]   S. Tehrani, E. Chen, M. Durlam, M. DeHerrera, J. M. Slaughter, J. Shi, and G. Kerszykowski. "High density submicron magnetoresistive random access memory (invited)". *Journal of Applied Physics*, Vol. 85, No. 8, pp. 5822–5827, Apr. 1999.

[Tser 02]   Y. Tserkovnyak, A. Brataas, and G. E. W. Bauer. "Enhanced Gilber damping in thin ferromagnetic films". *Physical Review Letters*, Vol. 88, No. 11, p. 117601, March 2002.

[Tsoi 00]   M. Tsoi, A. G. M. Jansen, J. Bass, W. C. Chiang, V. Tsoi, and P. Wyder. "Generation and detection of phase-coherent current-driven magnons in magnetic multilayers". *Nature*, Vol. 406, No. 6791, pp. 46–48, July 2000.

[Tsoi 98]   M. Tsoi, A. G. M. Jansen, J. Bass, W. C. Chiang, M. Seck, V. Tsoi, and P. Wyder. "Excitation of a magnetic multilayer by an electric current". *Physical Review Letters*, Vol. 80, No. 19, pp. 4281–4284, May 1998.

[Urba 01]   R. Urban, G. Woltersdorf, and B. Heinrich. "Gilbert damping in single and multilayer ultrathin films: Role of interfaces in nonlocal spin dynamics". *Physical Review Letters*, Vol. 87, No. 21, p. 217204, Nov. 2001.

[Van 00]   W. Van Roy, J. De Boeck, B. Brijs, and G. Borghs. "Epitaxial NiMnSb films on GaAs(001)". *Applied Physics Letters*, Vol. 77, No. 25, pp. 4190–4192, Dec. 2000.

[Van 01]   W. Van Roy, G. Borghs, and J. De Boeck. "Epitaxial growth of the half-metallic ferromagnet NiMnSb on GaAs(001)". *Journal Of Crystal Growth*, Vol. 227, pp. 862–866, July 2001.

[Van 02] W. Van Roy, G. Borghs, and J. De Boeck. "Epitaxial growth of NiMnSb on GaAs(111)A and B". *Journal of Magnetism and Magnetic Materials*, Vol. 242, pp. 489–491, Apr. 2002.

[Van 03] W. Van Roy, M. Wojcik, E. Jedryka, S. Nadolski, D. Jalabert, B. Brijs, G. Borghs, and J. De Boeck. "Very low chemical disorder in epitaxial NiMnSb films on GaAs(111)B". *Applied Physics Letters*, Vol. 83, No. 20, pp. 4214–4216, Nov. 2003.

[Walk 57] L. R. Walker. "Magnetostatic Modes In Ferromagnetic Resonance". *Physical Review*, Vol. 105, No. 2, pp. 390–399, 1957.

[Wijs 01] G. A. de Wijs and R. A. de Groot. "Towards 100NiMnSb/CdS interface". *Physical Review B*, Vol. 64, No. 2, p. 020402, July 2001.

[Zhou 06] Y. Zhou, C. McEvoy, R. Ramos, and I. V. Shvets. "The magnetic and magnetoresistance properties of ultrathin magnetite films grown on MgO substrate". *Journal of Applied Physics*, Vol. 99, No. 8, p. 08J111, Apr. 2006.

[Zhu 00] J. G. Zhu, Y. F. Zheng, and G. A. Prinz. "Ultrahigh density vertical magnetoresistive random access memory (invited)". *Journal of Applied Physics*, Vol. 87, No. 9, pp. 6668–6673, May 2000.

Acknowledgements

There are lots of people I would like to thank for their contribution to this work.

- I would like to thank Prof. Dr. Laurens W. Molenkamp for offering me the possibility to do a PhD at the "Lehrstuhl Experimentelle Physik III ".
- I would also like to thank Prof. Dr. Georg Schmidt who supervised me during my diploma thesis and partly through the PhD, he was always open for fruitful discussions.
- Dr. Charles Gould who supervised my PhD after Georg left to Halle and was always open for discussions.
- Thanks to Florian Lochner who was responsible for growing NiMnSb layers and special thanks to Bastian Büttner for the help and for just listen to problems I had.
- I would also like to thank Gabriel Dengel,Stefan Mark, Michael Rüth for helpful discussions and breakfast.
- A big thanks to Volkmar Hock. He was mostly helpful and always open for questions concerning technical issues.
- In the same manner I would like to mention Claus Schumacher with whom I spent hours at the XL30.
- And all the other people of the "Lehrstuhl Experimentelle Physik III ". Tanja Borzenko for discussions on lithography questions, Utz Bass for helping me with my car and Johannes Kleinlein who escaped to Halle but was always open for helpful discussions.
- Financial support from the European Union
- Last but not least I would like to thank my parents and Ann for always standing behind me and supporting me.

i want morebooks!

Buy your books fast and straightforward online - at one of world's fastest growing online book stores! Environmentally sound due to Print-on-Demand technologies.

Buy your books online at
www.get-morebooks.com

Kaufen Sie Ihre Bücher schnell und unkompliziert online – auf einer der am schnellsten wachsenden Buchhandelsplattformen weltweit! Dank Print-On-Demand umwelt- und ressourcenschonend produziert.

Bücher schneller online kaufen
www.morebooks.de

VDM Verlagsservicegesellschaft mbH
Heinrich-Böcking-Str. 6-8   Telefon: +49 681 3720 174   info@vdm-vsg.de
D - 66121 Saarbrücken   Telefax: +49 681 3720 1749   www.vdm-vsg.de

Printed by Books on Demand GmbH, Norderstedt / Germany